Stem Cell Biology and Regenerative Medicine

Series Editor
Kursad Turksen, Ph.D.
kursadturksen@gmail.com

More information about this series at http://www.springer.com/series/7896

Jeanne Wilson-Rawls • Kenro Kusumi
Editors

Innovations in Molecular Mechanisms and Tissue Engineering

Humana Press

Editors
Jeanne Wilson-Rawls
School of Life Sciences
Arizona State University
Tempe, AZ, USA

Kenro Kusumi
School of Life Sciences
Arizona State University
Tempe, AZ, USA

Neurogenomics Division
Translational Genomics Research Institute
Phoenix, AZ, USA

ISSN 2196-8985 ISSN 2196-8993 (electronic)
Stem Cell Biology and Regenerative Medicine
ISBN 978-3-319-44994-4 ISBN 978-3-319-44996-8 (eBook)
DOI 10.1007/978-3-319-44996-8

Library of Congress Control Number: 2016957175

Printed on acid-free paper

This Humana Press imprint is published by Springer Nature
The registered company is Springer International Publishing AG
The registered company address is: Gewerbestrasse 11, 6330 Cham, Switzerland

Preface

Interest in regeneration has waxed and waned since Lazzaro Spallanzani first described salamander limb regeneration in the eighteenth century. Currently, regeneration is a highly researched area with potential applications derived from biomedical and engineering research that will impact future medical therapies. The availability of genomic and transcriptomic data from regenerative species, combined with new approaches to identify and culture stem cells, has led to an explosive growth in our understanding of the molecular mechanisms of regeneration. In this book, we bring together the latest insights into these mechanisms.

In the first chapter, Debuque and Godwin describe the history of molecular research in salamanders and discuss the latest findings on unique proteins that mediate regeneration. These authors also provide insight into the regeneration of multiple structures and tissues and the transgenic tracing of stem cells that contribute to newly made tissues. While amphibians demonstrate spectacular regenerative ability, the more limited tail regeneration in anole lizards is captivating because these reptiles are evolutionarily more closely related to humans than salamanders, yet retain a significant ability to regenerate nerves, skin, muscle, and cartilage. In the second chapter, Hutchins et al. update the most recent findings in this model species.

The immune response is important for wound healing and initiation of regeneration. In the third chapter, Lynch and coauthors focus on the integration of the inflammatory response and the regulation of stem cells in regeneration of skeletal muscle in mammals. The role of the immune response is also discussed in amphibians and reptiles in the abovementioned chapters. Further, the immune response and signaling pathways during wound healing versus regeneration of amputated mouse digit tips is discussed by Dawson et al. This fourth chapter reviews important differences that account for regeneration following loss of the digit tip, but not more proximal amputations, as seen in very young children.

The next chapters focus on regeneration of cartilage, the heart, and the central nervous system. In the fifth chapter, Lozito et al. discuss cartilage regeneration in amphibians and reptiles and the application of findings from these species to human cartilage repair. Tissue engineering efforts focus on the repair of cartilage defects and of damage due to degeneration, such as in osteoarthritis. In the sixth chapter,

Judd and Huang provide a comprehensive discussion of cardiomyocyte regeneration in multiple model systems, including neonatal mice, with a focus on tissue engineering using stem cells. In the final chapter, Roussas et al. review traumatic brain injury and tissue engineering approaches to healing damage to the central nervous system.

In summary, this volume will appeal to readers interested in the broad overview of regenerative research, both in terms of species and tissues. Each chapter has a focus on molecular signals, the role of stem cells, and tissue engineering, making it a unique collection.

Tempe, AZ, USA Jeanne Wilson-Rawls
Tempe, AZ, USA Kenro Kusumi

Contents

Contributors

Caroline P. Addington School of Biological and Health Systems Engineering, Arizona State University, Tempe, AZ, USA

Alexander B. Andre School of Life Sciences, Arizona State University, Tempe, AZ, USA

Lindsay A. Dawson Department of Veterinary Physiology & Pharmacology, Texas A&M University, College Station, TX, USA

Ryan J. Debuque Australian Regenerative Medicine Institute, Monash University, Melbourne, VIC, Australia

Connor Dolan Department of Veterinary Physiology & Pharmacology, Texas A&M University, College Station, TX, USA

James W. Godwin Australian Regenerative Medicine Institute, Monash University, Melbourne, VIC, Australia

The Jackson Laboratory, Bar Harbor, ME, USA

The MDI Biological Laboratory, Bar Harbor, ME, USA

Guo N. Huang Department of Physiology, Cardiovascular Research Institute, University of California San Francisco, San Francisco, CA, USA

Elizabeth D. Hutchins School of Life Sciences, Arizona State University, Tempe, AZ, USA

Neurogenomics Division, Translational Genomics Research Institute, Phoenix, AZ, USA

Yangzi Jiang Department of Orthopaedic Surgery, Center for Cellular and Molecular Engineering, University of Pittsburgh School of Medicine, Pittsburgh, PA, USA

Justin Judd Cardiovascular Research Institute, University of California San Francisco, San Francisco, CA, USA

Paulina Ketcham Department of Veterinary Physiology & Pharmacology, Texas A&M University, College Station, TX, USA

Kenro Kusumi School of Life Sciences, Arizona State University, Tempe, AZ, USA

Neurogenomics Division, Translational Genomics Research Institute, Phoenix, AZ, USA

Tao Li Department of Hand Surgery, Wuhan Union Hospital, Tongji Medical College, Huazhong University of Science and Technology, Wuhan, Hubei, People's Republic of China

Thomas P. Lozito Department of Orthopaedic Surgery, Center for Cellular and Molecular Engineering, University of Pittsburgh School of Medicine, Pittsburgh, PA, USA

Cherie Alissa Lynch School of Life Sciences, Arizona State University, Tempe, AZ, USA

Briana I. Martinez School of Biological and Health Systems Engineering, Arizona State University, Tempe, AZ, USA

Ken Muneoka Department of Veterinary Physiology & Pharmacology, Texas A&M University, College Station, TX, USA

Alan Rawls School of Life Sciences, Arizona State University, Tempe, AZ, USA

Adam Roussas School of Biological and Health Systems Engineering, Arizona State University, Tempe, AZ, USA

Paula P. Schanes Department of Veterinary Physiology & Pharmacology, Texas A&M University, College Station, TX, USA

Sarah E. Stabenfeldt School of Biological and Health Systems Engineering, Arizona State University, Tempe, AZ, USA

Rocky S. Tuan Department of Orthopaedic Surgery, Center for Cellular and Molecular Engineering, University of Pittsburgh School of Medicine, Pittsburgh, PA, USA

Jeanne Wilson-Rawls School of Life Sciences, Arizona State University, Tempe, AZ, USA

Mingquan Yan Department of Veterinary Physiology & Pharmacology, Texas A&M University, College Station, TX, USA

Ling Yu Department of Veterinary Physiology & Pharmacology, Texas A&M University, College Station, TX, USA

Chapter 1
Research into the Cellular and Molecular Mechanisms of Regeneration in Salamanders: Then and Now

Ryan J. Debuque and James W. Godwin

1.1 Introduction

Regenerative medicine encompasses collaboration between scientists with diverse backgrounds in wound healing, immunology, developmental biology, stem cell science, tissue engineering and more recently, organic chemistry and nanotechnology. Research in this area aims to improve patient outcomes in the contexts of chronic diseases, ageing and acute injuries. Therapeutic efforts have focused on delivering single molecules, embryonic or adult stem cell derived tissues supported by artificial scaffolds to either directly replenish lost tissue or provide paracrine factors to enhance local wound healing [1]. Promising studies in mammalian in vitro and in vivo models following this strategy have spawned development of numerous clinical trials to varying levels of success but have yet to emulate major aspects of true regeneration exemplified in nature.

Experimental biologists have been studying natural adult regeneration in many phyla for centuries and include vertebrates such as amphibians and teleosts reviewed in [2–6]. Urodele amphibians (commonly referred to as salamanders) are some of the oldest animals to be housed in laboratories and have contributed to many fundamental concepts and discoveries in experimental biology [7]. Regeneration is a property shared in all ten families but is best understood in a few species representing selected genera [8]. Evolutionary reasons for the robust

R.J. Debuque
Australian Regenerative Medicine Institute, Monash University, Melbourne, VIC, Australia

J.W. Godwin (✉)
Australian Regenerative Medicine Institute, Monash University, Melbourne, VIC, Australia

The Jackson Laboratory, Bar Harbor, ME, USA

The MDI Biological Laboratory, Bar Harbor, ME, USA
e-mail: james.godwin@jax.org

J. Wilson-Rawls, K. Kusumi (eds.), *Innovations in Molecular Mechanisms and Tissue Engineering*, Stem Cell Biology and Regenerative Medicine,
DOI 10.1007/978-3-319-44996-8_1

regenerative potential in salamanders, in comparison to mammals is poorly understood. Classically it has been hypothesized that mammals have lost the cellular machinery or processes required for *scarless* healing, a potential requirement for perfect tissue regeneration, possibly in favor of selecting for strategies to deal with a wider range disease pathogens [9–11]. An alternative reason centers around the idea that any species with the capacity to regenerate complex tissues may have acquired this ability through the selection of specific genes and is not an ancestrally shared property [12].

Regeneration in salamanders was first documented in 1776 by the Italian scientist Lazzaro Spallanzani who documented the regeneration of the forelimb, hindlimb, tail, gill and jaw [13]. This chapter encapsulates the major discoveries in the 240 years since, covering fundamental ideas originating from surgical manipulations, insights aided by molecular tools and potential outcomes anticipated using next generation sequencing and genome editing technologies.

1.2 Establishing the Tissue Requirements and Boundaries for Regeneration

1.2.1 The Regenerative Potential of Salamanders

Early inquiries into the regenerative potential in salamanders defined the limb, heart, brain, lens, tail, spinal cord, liver, jaw, bone segments, muscle, skin, and gills as tissues capable of growth after resection (See Table 1.1) [13, 78, 85, 102, 120, 135]. Studies concerning the regeneration of clinically relevant tissues (heart, spinal

Table 1.1 Regenerating tissues in salamanders

Tissue	Primary research references
Limb	[14–54][a] [18, 28, 39, 55–65][b] [66–68][c] [69–71][d] [72–75][e]
Heart	[76, 77][a] [78–84][b]
Brain	[85, 86][a] [87–90][b]
Lens/ Retina	[91][a] [92–96][b] [97][c] [98–101][e]
Spinal cord	[102–111][a] [112–115][b] [116][d] [117]
Liver	[118, 119][e]
Jaw	[120–122][b] [123][c] [124][d]
Bone	[125, 126][a] [127][c]
Muscle	[128–131][a] [131–134][b]

[a]Ambystoma
[b]Notopthalamus
[c]Cynops
[d]Pleurodeles
[e]Triturus

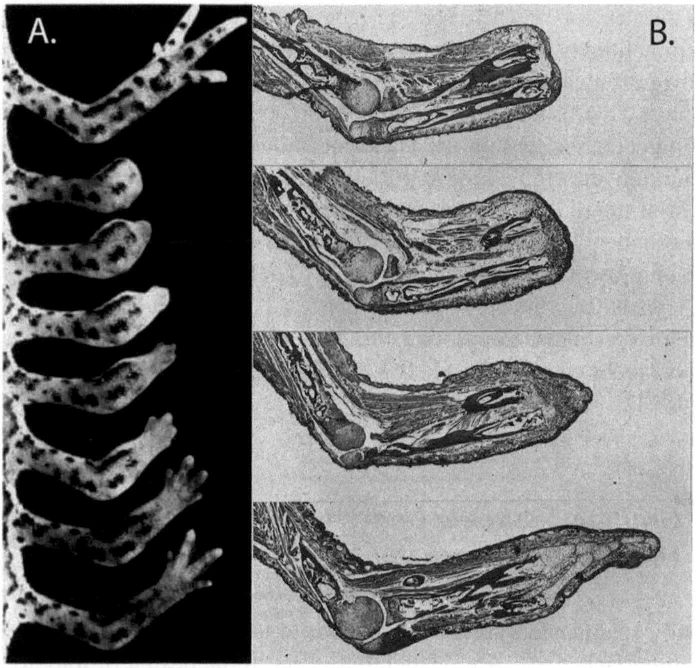

Fig. 1.1 Morphological and histological view of salamander limb regeneration. (**a**) Gross anatomical view of the successive stages of regeneration in the newt. (**b**) Histological sections of regenerating newt limbs. Following amputation, cells of the epidermis have migrated over to cover the wound and thicken to form a structure known as the wound epithelium (WE) or apical epidermal cap (AEC) (First top two images). Interactions between the WE and nerve provide mitogenic signals to cells beneath the WE to initiate a proliferative response resulting in the formation of a blastema (third image from the top). Proximal blastema cells differentiate and develop into new limb tissues such as bone, muscle and nerve to restore normal limb architecture. Adapted from [136]

cord, brain, and lens) were reported less frequently compared to the limb. Adult salamander limb regeneration is a unique property not observed in any other tetrapod. The limb is very amenable to complex procedures with low mortality risks and shares structural similarities to mammals. The process of limb regeneration after amputation has been well characterized and initially defined from early gross anatomical and histological observations (Fig. 1.1).

1.2.2 Tissue Requirements for Limb Regeneration

Many of the experiments performed during this era aimed at identifying methods to prevent or perturb limb regeneration. Loss of function studies carried out by scientists at the time utilized two main methods, surgical removal of specific tissues in the limb or ablation of blastema cells via irradiation [14]. Key findings using these methods were the identification of the nerve and wound epithelium as essential tissues for regeneration. The clearest example for the requirement of wound epithelium came from experiments demonstrating a blockade of limb outgrowth by grafting a flap of intact skin over the site of amputation [15, 16]. Failure of this outgrowth was attributed to a reduction in cellular proliferation after the first week of regeneration, within the blastema (mound of progenitor cells forming at the amputation site) [55]. First reported in 1823, de-nervation of the limb either prior to or at the time of amputation results in the formation of a scar-less stump [137]. Subsequent studies both in the salamander and anuran amphibians identified that limb outgrowth is dependent on density of nerve tissue, not type of innervation and that signals from the nerve control blastema outgrowth [17, 56, 72, 138–141]. Additional experiments supporting this idea originated from experiments where nerves were resected and deviated towards foreign areas to produce supernumerary limbs [74, 142, 143].

1.2.3 Grafting Tissues to Understand Positional Identity During Limb Regeneration

Historically salamanders have been known to tolerate both allografts and xenografts without acute rejection, which has allowed the design of long term regeneration studies featuring tissue grafts [144, 145]. In particular this technique has been useful for understanding ideas regarding positional identity and memory during regeneration of a tissue. In the case of the limb, regeneration occurs across three dimensional axis (proximal-distal, anterior-posterior and dorsal-ventral). Most experiments examining positional memory have looked at the proximal-distal axis (shoulder-wrist). One example is the experiment performed by Goss, who implanted a distal amputated limb into the flank after which resection of the elbow joint (originally proximal) displayed outgrowth of distal skeletal elements (wrist) [146].

Another example was the finding that intercalary regeneration (replacement of missing structures between two juxtaposed tissues) is unidirectional and proceeds in a proximal-distal fashion (referred to as the law of distal transformation) [19, 57]. Other approaches to studying positional identity involved the use of grafting blastemas from different levels along the PD axis onto the dorsal side of proximal stumps to observe the displacement of the grafted tissue back to its original position and then proceeding with limb outgrowth [20].

Further work using tissue-grafting experiments established the concept of positional discontinuity during the early stages of regeneration as a requirement for outgrowth. Originating from studies in invertebrate models, positional discontinuity is achieved when tissues from opposite sides of an axis confront each other (e.g. dermis from the anterior side of an amputated limb meets with the posterior side) [147]. Experiments focusing on the relationship of cells along transverse axes of the limb (anterior-posterior and dorsal-ventral) demonstrated this requirement by inducing supernumary tissues to form by rotating tissues of a stump following amputation or rotating a blastema and grafting them to a stump [21, 22, 57, 58, 148]

1.3 Molecular Mechanisms and Cellular Dynamics of Regeneration

1.3.1 Identifying the Molecular Mechanisms Underlying Limb Regeneration

Experimental approaches for dissecting molecules that regulate limb regeneration were inspired by research conducted in the late 1970s by Niazi and Saxena who first reported the abnormal effects of vitamin A on limb regeneration in tadpoles [149]. Repeated in the axolotl shortly after, Maden was able to show that retinoic acid and its derivatives were able to reject the law of distal transformation and cause proximal limb elements to regenerate from a distal amputation [23]. Subsequent studies later found that regeneration along the transverse axis of the limb was also perturbed and have implicated additional roles for retinoic acid signaling in other regenerating tissues [18, 150, 151].

Research spawning from the influence of retinoic acid aimed to utilise the molecular tools of the early 1990s to elucidate roles for candidate genes regulating limb regeneration. Inspiration for choosing candidates to examine came from a plethora of studies on vertebrate limb development, which had well defined morphogenetic signalling zones. Blastema outgrowth and patterning shares many structural similarities to a developing limb thus it is logical to assume that the same molecules have similar roles. Indeed such a hypothesis is supported with several studies elucidating roles or identifying expression patterns of genes belonging to several developmental signalling pathways such as Hox, Fgf, Hh, Bmp and Wnt [25–27, 68].

A molecular explanation for retinoic acid's control across the PD axis came with the identification of Prod1 [59]. Identified in a subtractive cDNA screen of cultured newt blastema cells, Prod1 is known to be expressed at the cell surface and regulated by retinoic acid and Meis homeoprotein during limb regeneration [29, 30, 59]. It is one of the few salamander proteins to have its structure solved and is present in nine salamander species spanning four families [152, 153]. Interestingly this gene is required for pre-axial digit formation and has no known mammalian orthologues making it one of the few known salamander specific genes involved in limb regeneration [12, 28]. Prod1 is also indirectly involved in nerve dependent regeneration where it has been shown to bind to the newt orthologue/paralogue of anterior gradient protein 2 (nAG) [60]. nAG is expressed first at severed nerve sheaths, secreted by Schwann cells and subsequently in gland cells of the wound epithelium. Over-expression of this protein is sufficient to stimulate blastema cell outgrowth of de-nervated limbs, rescue limb regeneration, and provide novel a molecular pathway to study nerve dependent regeneration [60].

1.3.2 New Insights to Cellular Contributions During Limb Regeneration

Advancements in imaging in cloning have allowed the production of genetic tools to fluorescently label whole animals or tissues and visualise cellular dynamics during regeneration [154]. Grafting fluorescent donor tissues into wild-type is one of the most common strategies applied when studying cellular contribution during regeneration, a feature accessible to few models [31, 32, 130, 131]. One study to take advantage of this strategy is the landmark paper by Kragl and colleagues who determined the heterogeneity of cells in the blastema and their restricted nature to contribute to tissues differing from their embryonic origin [33]. Another example was the deployment of a suite of HOXA antibodies staining donor GFP connective tissue blastema cells during limb regeneration, overturning fundamental concepts regarding segment formation along the proximal distal axis [34–36].

Research into the potential contributions of the immune system to regeneration has historically been limited with most studies examining its relationship in the wound healing response. Clear documentation of infiltrating leukocyte kinetics and the influences of peripheral lymphoid organs was reported in the 1980s [61, 155]. Many immune-modulating drugs and procedures have been screened for effects on outgrowth with other studies implicating roles for known mammalian genes involved with wound healing [37, 62–64]. Renewed efforts in this area have seen the development and adaptation of modern techniques to study the immune system in the salamander [156, 157]. Recent evidence using the axolotl suggests that cells of the evolutionarily conserved innate immune system, particularly macrophages have roles through all phases of regeneration [38]. Macrophage depletion after blastema formation allows regeneration to complete, but is delayed. Macrophage ablation prior to amputation has been shown to block limb regeneration but re-amputation following replenishment resets normal limb regeneration indicating a temporal requirement for these cells prior to blastema formation [38]. Failed limb regenerations are characterised by collagen rich scars, decreased cellular proliferation, and alterations to key regeneration associated genes such as MMPs, and TGF-β [37, 38, 65]. Further work also implicated a role for dependent immune-surveillance and clearance of senescent cells in the regenerating limb [39].

1.3.3 *Mechanisms of Outgrowth During Regeneration Found to be Tissue Specific*

A key concept of regenerative medicine is that mechanisms governing cellular outgrowth are not universal and should be considered tissue specific (Fig. 1.2). Generally speaking, replacement of a new structure requires cells to arise and undergo several rounds of rapid proliferation. The mode through which this is conducted can be through the recruitment and activation of stem-progenitor cells or stimulating resident and neighbouring post-mitotic cells to re-enter the cell-cycle. One example of the former is regeneration of the spinal cord. Regeneration of the spinal cord following resection activates resident neural stem cells to mobilise locally and then migrate along the anterior-posterior axis. These cells serve as a multi-potent source for all neural cells in the regenerated tissue [103–105]. Molecules implicated for regulating in this process include planar cell polarity genes and microRNAs both conserved and unique to the salamander

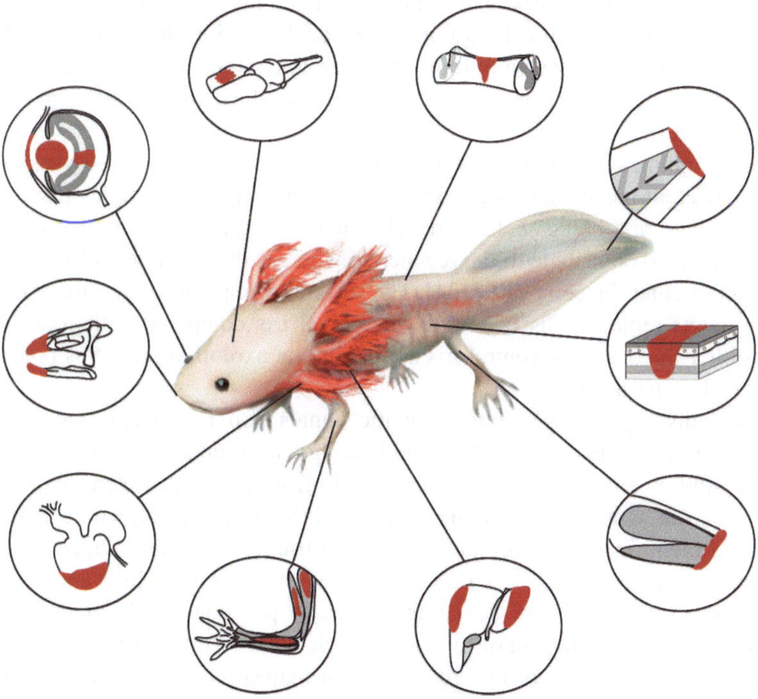

Fig. 1.2 A range of clinically relevant tissues can be regenerated in adult salamanders. This figure illustrates a generalized summary of regenerating tissue in salamanders using various species as outlined in Table 1.1. The tissues identified so far include: brain, spinal cord, tail, skin, limbs, liver, skeletal muscle, heart, jaws, and ocular tissues such as retina, cornea, and lens. Variation in modes of regeneration and adult capacity are outlined in the text. Injury site is highlighted in *red*. Axolotl image provided and adapted with permission from Memuco© artist services and IUCN Arkive

[106, 107, 109, 158]. Similar mechanisms have been observed in the case of the regenerating telencephalon and dopaminergic neurons of the mid-brain. Following mechanical removal or chemical ablation, cells within these tissues have been shown to cause rapid proliferation and neurogenesis from spatio-temporal restricted zones [86–90].

In contrast many studies have reported that the salamander lens and heart utilise the second mode of replacing cells, which interestingly has limited examples in mammalian regeneration [76, 79, 92]. One of the most studied areas of salamander biology is lens regeneration. After lentectomy, pigment epithelial cells originating from the dorsal iris, re-enter the cell cycle, lose their pigmentation and other differentiated characteristics, before undergoing trans-differentiation into new lens tissue. This trans-differentiation is accompanied by the activation of sequential lens development gene expression, reviewed in detail elsewhere [159]. It should be noted that this process can be repeated almost indefinitely as mounting evidence from both histological and molecular studies suggests that lens regeneration is not affected by age or the number of times it is removed [93, 94]. Similarly cardiomyocytes can lose many of their differentiated characteristics and proliferate following ventricle resection, replacing up to 20 % of the original ventricle tissue [76, 80]. Signals initiating cell-cycle re-entry have yet to be identified, however one known important regulator is components of the extracellular matrix, which has shown to undergo rapid changes during the early stages of heart regeneration [80–82].

An intriguing aspect to keep in mind is the potential for identical tissues to make use of different mechanisms between species. Already two examples for this have emerged. The first being skeletal muscle where axolotls deploy activated resident satellite cells to contribute to the regenerate whereas the myofibers of the red spotted newt re-enter the cell cycle [131]. The second example is in the case of the lens, where newts replace cells from only the dorsal iris compared with contributions from either the dorsal or ventral iris as seen in the axolotl, though this potential is lost shortly after hatching [91].

Distinguishing the regeneration specific signals from the background noise arising from amputation associated wound healing and trauma, is extremely difficult. Ideally studies elucidating molecular signals from essential regenerative tissues (e.g. nerve or wound epithelium) should reduce irrelevant signaling that could mask the identification of key pathways and obscure accurate interpretation. One available assay that addresses these criteria is the accessory limb model, which produces ectopic limbs by deviating nerves to positionally discontinuous skin grafts [40]. This unique gain of function ectopic outgrowth assay in the salamander is an extremely useful tool in a model where majority of functional experiments involve loss of function studies. Indeed several molecules have been tested in this system and should gain future utility testing novel candidate genes required for limb regeneration [41–44].

1.4 Entering the era of Next Generation Sequencing and Genome Editing

1.4.1 Unraveling the Salamander Genome and High-Throughput Sequencing Studies in Regeneration

Salamanders are known to have some of the largest genomes amongst all vertebrates with some species approximately containing between 14 and 120 Gb, compared with a genome size of around 3.2 Gb in humans [160, 161]. Characterized by high percentages of transposable elements, the genomic gigantism observed across the salamander family are hypothesized to have originated from a shared period of genome expansion during the Jurassic era [162, 163]. As such, complete genome assemblies are lacking in all families, which is perhaps the biggest drawback for any prospective academics interested in working with the model. Several resources have been put in place to obtain sequence information, largely derived from transcriptome and proteomic analysis from tissues across multiple species [45, 46, 83, 161, 164]. In addition online repositories are available and regularly updated with omics data from the latest studies [165–167].

Many experiments have already utilized next generation sequencing technologies for high through-put transcriptome analysis during limb, spinal cord and lens regeneration [47–49, 95, 110]. One example was the time course analysis performed by Knapp and colleagues examining the transcriptional changes over the course of limb regeneration [45]. This approach revealed that gene expression follows a similar pattern as seen in morphological studies with signature wound healing genes first among those upregulated, followed by amputation associated regenerative genes and then finally, genes implicated in limb development [45]. Studies utilizing proteomics have also been conducted across multiple tissues and species and have contributed major findings such as the identification of novel newt specific CCN, a protein located in the endocardium that is specifically upregulated during the early stages of heart regeneration [83, 168, 169].

Ultimately sequenced based inquiry into the genetic networks of regeneration will require a complete genome assembly however progress towards obtaining genomic information in any species has been understandably limited. Encouragingly the first characterization of the axolotl genome has been documented [170]. Estimated to be 32 Gb in size, the axolotl genome provides an example of the difficulties associated with assembling sequences from large genomes as well as the potential approaches used to overcome current computational limitations [170].

1.4.2 Genome Editing Technologies

Innovations over the last decade in the field of molecular biology have provided a multitude of options to genetically modify the salamander Table 1.2. The application of Sce1-meganuclease or Tol2-transposase technology has produced germline transgenics expressing ubiquitous fluorescent proteins in various salamander species [154, 175, 176, 179]. Several transgenic reporters have been developed to track the fates of tissues such as nerve, Schwann cells, muscle, epidermis and cartilage or signaling molecules like retinoic acid [51, 52]. In addition it is now possible to temporally control gene expression within specific cells thanks to the development of

Table 1.2 Genetic tools available in salamanders

Genome editing tool		References
Non-germline vectors		
Viruses	Vaccinia virus	[171][a]
	Adenovirus	[133, 172][b]
	Pseudotyped virus	[173][a]
	Foamy virus	[174][a]
Germline transgenics		
Ubiquitous reporters	CAGGS;EGFP	[154, 175, 176][a,c,d]
	CAGGS;CherryNuc	[33][a]
	CAGGS;LP-EGFP-LP-Tomato	[51][a]
	CAGGS;LP-EGFP-LP-p16-T2A-Cherry	[51][a]
	CAGGS;ER-Cre-ER-T2A-EGFP-nuc	[51][a]
Tissue specific reporters	B3Tubulin:EGFP	[51][a]
	CNP;EGFP	[51][a]
	Col2a1:EGFP	[51][a]
	Krt12:EGFP	[51][a]
	CarAct;EGFP	[51][a]
	AxSox2;cre-ert2-T2A-GFP	[51][a]
	Col2A1:ER-Cre-ER-T2A-EGFP-nuc	[51][a]
Signaling molecule reporters	RARE;EGFP	[52][a]
Loss of function genetics		
TALENs	Tryosinase	[177][d]
	Sox2	[111][a]
	Thrombospondin-1	[54][a]
	Prod1	[28][b]
CRISPR	Brachyury	[178][a]
	Sox2	[111][a]

[a]Ambystoma
[b]Notophthalamus
[c]Cynops
[d]Pleurodeles

loxP and Cre-driver lines that can be bred together and supplemented with tamoxifen to induce Cre-mediated recombination [51]. Such technologies have already allowed fate mapping studies in the axolotl to be performed and have provided the tools necessary to design complex experiments, which have yielded evidence for species specific mechanisms of tissue regeneration [131]. It should be noted that considerable time is required for germline transgenesis to occur. Thus targeting specific cell types can be achieved in mosaic backgrounds by delivering vectors such as vaccinia virus, adenovirus, pseudotyped virus, and foamy virus to allow more rapid analysis of phenotypes [133, 171, 173, 174].

Traditionally tools to perform loss of function genetics during regeneration in the salamander were limited to the use of morpholinos, which have potential to cause off-target effects [53, 96, 180]. Alternative protocols to perturb gene function have focused on inducing double-strand breaks, which often leave insertions or deletions following non-homologous end joining repair. Two of these methods; transactivator-like effector nucleases (TALENs) and clustered regularly interspaced short palindromic repeat (CRISPR) have already completed proof of concept studies in the axolotl, red-spotted and Iberian ribbed newt (*Pleurodeles waltl*) [28, 54, 111, 177, 178].

1.5 Conclusions

1.5.1 The Influence of Regeneration Research in Salamanders

Many underlying concepts concerning the determinants for successful regeneration have arisen from research in salamanders. Understanding how regenerating cells can dynamically navigate across a three-dimensional axis and recapture its original form is critical in developing complex tissue transplantation models. Innervation and the supply of neurotrophic signals to the tissues of the regenerate has been explored and implicated in both teleost and mammalians models of regeneration including the heart, digit tip, earlobe, bone marrow, and hair follicles [181–186]. The resolution of wound healing and the subsequent transition to cell cycle re-entry requires precise co-ordination between infiltrating immune cells and the local tissue environment harboring resident progenitor-stem populations. Evidence for this emerging theme is springing from the salamander, anuran amphibian, and teleost systems, which has influenced transitional studies in most mammalian models of regeneration [187–191].

1.5.2 Future Perspectives

Considerable advances have been made in recent years to improve genomic and molecular resources in all salamander species that are regularly used in the laboratory. These technologies have enabled scientists to revisit classic experiments with greater resolution to overturn or confirm fundamental ideas as well as develop new lines of investigation to pursue. Progress towards complete genome assemblies will be a challenging but is an essential resource for future studies.

Developing new genetic tools to follow specific cellular movements, interactions and contributions in all regenerating tissues is necessary to drive the model forward. Particular focus will evolve towards identifying taxon-specific genes or molecules with known orthologues through next-generation sequencing or candidate based approaches. Emphasis will be placed on combining knock-down genetics with gain of function assays unique to the salamander system to define molecular mechanisms. Research efforts in this area will enable the development of in vitro and in vivo gain of function assays in mammals with the eventual goal of translating these findings for the treatment of human diseases and injuries.

References

1. Forbes SJ, Rosenthal N (2014) Preparing the ground for tissue regeneration: from mechanism to therapy. Nat Med 20:857–869. doi:10.1038/nm.3653
2. Goss RJ (2013) Principles of regeneration. Academic, New York
3. Dinsmore CE (2007) A history of regeneration research. Cambridge University Press, Cambridge
4. Tanaka EM, Reddien PW (2011) The cellular basis for animal regeneration. Dev Cell 21:172–185. doi:10.1016/j.devcel.2011.06.016
5. Carlson BM (2011) Principles of regenerative biology. Academic, Burlington, MA
6. Stocum DL (2012) Regenerative Biology and Medicine, 2nd edn. Academic, San Diego
7. Reiß C, Olsson L, Hoßfeld U (2015) The history of the oldest self-sustaining laboratory animal: 150 years of axolotl research. J Exp Zool B Mol Dev Evol 324:393–404. doi:10.1002/jez.b.22617
8. Kumar A, Simon A (2015) Salamanders in regeneration research. Humana Press, New York, 10.1007/978-1-4939-2495-0
9. Fröbisch NB, Bickelmann C, Witzmann F (2014) Early evolution of limb regeneration in tetrapods: evidence from a 300-million-year-old amphibian. Proc Biol Sci 281:20141550. doi:10.1098/rspb.2014.1550
10. Godwin JW, Brockes JP (2006) Regeneration, tissue injury and the immune response. J Anat 209:423–432. doi:10.1111/j.1469-7580.2006.00626.x
11. Godwin JW, Rosenthal N (2014) Scar free wound healing in amphibians: immunological influences on regenerative success. Differentiation 87:66–75. doi:10.1016/j.diff.2014.02.002
12. Garza-Garcia AA, Driscoll PC, Brockes JP (2010) Evidence for the local evolution of mechanisms underlying limb regeneration in salamanders. Integr Comp Biol 50:528–535. doi:10.1093/icb/icq022
13. Spallanzani L (1769) An essay on animal reproductions. In: Spallanzani L (ed) An essay on animal reproductions (translated from the italian). Becket and Hondt, London, pp 68–82
14. Thornton CS (1958) The inhibition of limb regeneration in urodele larvae by localized irradiation with ultraviolet light. J Exp Zool 137:153–179. doi:10.1002/jez.1401370108
15. Tornier G (1906) Kampf der Gewebe im Regenerat bei Begünstigung der Hautregeneration. Archiv für Entwicklungsmechanik der Organismen 22:348–369. doi:10.1007/BF02162806
16. Godlewski E (1928) Untersuchungen über Auslösung und Hemmung der Regeneration beim Axolotl. Wilhelm Roux Arch Entwickl Mech Org 114:108–143
17. Thornton CS (1954) The relation of epidermal innervation to limb regeneration in *Ambystoma* larvae. J Exp Zool 127:577–601. doi:10.1002/jez.1401270307

18. Stocum DL, Thoms SD (1984) Retinoic-acid-induced pattern completion in regenerating double anterior limbs of urodeles. J Exp Zool 232:207–215. doi:10.1002/jez.1402320208
19. Stocum DL (1975) Regulation after proximal or distal transposition of limb regeneration blastemas and determination of the proximal boundary of the regenerate. Dev Biol 45:112–136. doi:10.1016/0012-1606(75)90246-8
20. Crawford K, Stocum DL (1988) Retinoic acid coordinately proximalizes regenerate pattern and blastema differential affinity in axolotl limbs. Development 102:687–698
21. Carlson BM (1974) Morphogenetic interactions between rotated skin cuffs and underlying stump tissues in regenerating axolotl forelimbs. Dev Biol 39:263–285. doi:10.1016/S0012-1606(74)80029-1
22. Carlson BM (1975) The effects of rotation and positional change of stump tissues upon morphogenesis of the regenerating axolotl limb. Dev Biol 47:269–291. doi:10.1016/0012-1606(75)90282-1
23. Maden M (1982) Vitamin A and pattern formation in the regenerating limb. Nature 295:672–675. doi:10.1038/295672a0
24. Kim W-S, Stocum DL (1986) Effects of retinoids on regenerating limbs: comparison of retinoic acid and arotinoid at different amputation levels. Rouxs Arch Dev Biol 195:455–463. doi:10.1007/BF00375749
25. Gardiner DM, Blumberg B, Komine Y, Bryant SV (1995) Regulation of HoxA expression in developing and regenerating axolotl limbs. Development 121:1731–1741
26. Mullen LM, Bryant SV, Torok MA et al (1996) Nerve dependency of regeneration: the role of Distal-less and FGF signaling in amphibian limb regeneration. Development 122:3487–3497. doi:10.1016/0012-1606(92)90047-K
27. Kawakami Y, Rodriguez Esteban C, Raya M et al (2006) Wnt/beta-catenin signaling regulates vertebrate limb regeneration. Genes Dev 20:3232–3237. doi:10.1101/gad.1475106
28. Kumar A, Gates PB, Czarkwiani A, Brockes JP (2015) An orphan gene is necessary for preaxial digit formation during salamander limb development. Nat Commun 6:8684. doi:10.1038/ncomms9684
29. Mercader N (2005) Proximodistal identity during vertebrate limb regeneration is regulated by Meis homeodomain proteins. Development 132:4131–4142. doi:10.1242/dev.01976
30. Shaikh N, Gates PB, Brockes JP (2011) The Meis homeoprotein regulates the axolotl Prod 1 promoter during limb regeneration. Gene 484:69–74. doi:10.1016/j.gene.2011.06.003
31. McCusker CD, Gardiner DM (2013) Positional information is reprogrammed in blastema cells of the regenerating limb of the axolotl (*Ambystoma mexicanum*). PLoS One 8, e77064. doi:10.1371/journal.pone.0077064.s005
32. Nacu E, Glausch M, Le HQ et al (2013) Connective tissue cells, but not muscle cells, are involved in establishing the proximo-distal outcome of limb regeneration in the axolotl. Development 140:513–518. doi:10.1242/dev.081752
33. Kragl M, Knapp D, Nacu E et al (2009) Cells keep a memory of their tissue origin during axolotl limb regeneration. Nature 460:60–65. doi:10.1038/nature08152
34. Faber J (1960) An experimental analysis of regional organization in the regenerating fore limb of the axolotl (*Ambystoma mexicanum*). Arch Biol (Liege) 71:1–72
35. Echeverri K, Tanaka EM (2005) Proximodistal patterning during limb regeneration. Dev Biol 279:391–401. doi:10.1016/j.ydbio.2004.12.029
36. Roensch K, Tazaki A, Chara O, Tanaka EM (2013) Progressive specification rather than intercalation of segments during limb regeneration. Science 342:1375–1379. doi:10.1126/science.1241796
37. Lévesque M, Gatien S, Finnson K et al (2007) Transforming growth factor β signaling is essential for limb regeneration in axolotls. PLoS One 2, e1227. doi:10.1371/journal.pone.0001227.t001
38. Godwin JW, Pinto AR, Rosenthal NA (2013) Macrophages are required for adult salamander limb regeneration. Proc Natl Acad Sci U S A 110:9415–9420. doi:10.1073/pnas.1300290110/-/DCSupplemental/pnas.201300290SI.pdf
39. Yun MH, Davaapil H, Brockes JP (2015) Recurrent turnover of senescent cells during regeneration of a complex structure. Elife 4, e05505. doi:10.7554/eLife.05505

40. Endo T, Bryant SV, Gardiner DM (2004) A stepwise model system for limb regeneration. Dev Biol 270:135–145. doi:10.1016/j.ydbio.2004.02.016
41. Satoh A, Gardiner DM, Bryant SV, Endo T (2007) Nerve-induced ectopic limb blastemas in the axolotl are equivalent to amputation-induced blastemas. Dev Biol 312:231–244. doi:10.1016/j.ydbio.2007.09.021
42. Satoh A, Hirata A, Makanae A (2012) Collagen reconstitution is inversely correlated with induction of limb regeneration in *Ambystoma mexicanum*. Zoolog Sci 29:191–197. doi:10.2108/zsj.29.191
43. Makanae A, Hirata A, Honjo Y et al (2013) Nerve independent limb induction in axolotls. Dev Biol 381(1):1–14. doi:10.1016/j.ydbio.2013.05.010
44. Makanae A, Mitogawa K, Satoh A (2014) Co-operative Bmp- and Fgf-signaling inputs convert skin wound healing to limb formation in urodele amphibians. Dev Biol 396:57–66. doi:10.1016/j.ydbio.2014.09.021
45. Knapp D, Schulz H, Rascon CA et al (2013) Comparative transcriptional profiling of the axolotl limb identifies a tripartite regeneration-specific gene program. PLoS One 8, e61352. doi:10.1371/journal.pone.0061352.s011
46. Stewart R, Rascon CA, Tian S et al (2013) Comparative RNA-seq analysis in the unsequenced axolotl: the oncogene burst highlights early gene expression in the blastema. PLoS Comput Biol 9, e1002936. doi:10.1371/journal.pcbi.1002936.s012
47. Monaghan JR, Athippozhy A, Seifert AW et al (2012) Gene expression patterns specific to the regenerating limb of the Mexican axolotl. Biol Open 1:937–948. doi:10.1242/bio.20121594
48. Holman EC, Campbell LJ, Hines J, Crews CM (2012) Microarray analysis of microrna expression during axolotl limb regeneration. PLoS One 7, e41804. doi:10.1371/journal.pone.0041804
49. Wu C-H, Tsai M-H, Ho C-C et al (2013) De novo transcriptome sequencing of axolotl blastema for identification of differentially expressed genes during limb regeneration. BMC Genomics 14:434. doi:10.1186/1471-2164-14-434
50. Rao N, Jhamb D, Milner DJ et al (2009) Proteomic analysis of blastema formation in regenerating axolotl limbs. BMC Biol 7:83. doi:10.1186/1741-7007-7-83
51. Khattak S, Schuez M, Richter T et al (2013) Germline transgenic methods for tracking cells and testing gene function during regeneration in the axolotl. Stem Cell Rep 1:90–103. doi:10.1016/j.stemcr.2013.03.002
52. Monaghan JR, Maden M (2012) Gene expression patterns specific to the regenerating limb of the Mexican axolotl. Dev Biol 368:63–75. doi:10.1016/j.ydbio.2012.05.015
53. Zhu W, Pao GM, Satoh A et al (2012) Activation of germline-specific genes is required for limb regeneration in the Mexican axolotl. Dev Biol 370:42–51. doi:10.1016/j.ydbio.2012.07.021
54. Kuo TH, Kowalko JE, DiTommaso T et al (2015) TALEN-mediated gene editing of the thrombospondin-1 locus in axolotl. Regeneration 2:37–43. doi:10.1002/reg2.29
55. Mescher AL (1976) Effects on adult newt limb regeneration of partial and complete skin flaps over the amputation surface. J Exp Zool 195:117–127. doi:10.1002/jez.1401950111
56. Brockes JP, Kintner CR (1986) Glial growth factor and nerve-dependent proliferation in the regeneration blastema of Urodele amphibians. Cell 45:301–306
57. Iten LE, Bryant SV (1975) The interaction between the blastema and stump in the establishment of the anterior-posterior and proximal-distal organization of the limb regenerate. Dev Biol 44:119–147. doi:10.1016/0012-1606(75)90381-4
58. Bryant SV, Iten LE (1976) Supernumerary limbs in amphibians: experimental production in *Notophthalmus viridescens* and a new interpretation of their formation. Dev Biol 50:212–234. doi:10.1016/0012-1606(76)90079-8
59. da Silva SM, Gates PB, Brockes JP (2002) The newt ortholog of cd59 is implicated in proximodistal identity during amphibian limb regeneration. Dev Cell 3:547–555. doi:10.1016/S1534-5807(02)00288-5
60. Kumar A, Godwin JW, Gates PB et al (2007) Molecular basis for the nerve dependence of limb regeneration in an adult vertebrate. Science 318:772–777. doi:10.1126/science.1147710

61. Fini ME, Sicard RE (1980) Limb regeneration of the adult newt (*Notophthalmus viridescens*) in the absence of the spleen. Rouxs Arch Dev Biol 189:77–79. doi:10.1007/BF00848570
62. Schotté OE, Sicard RE (1982) Cyclophosphamide-induced leukopenia and suppression of limb regeneration in the adult newt, *Notophthalmus viridescens*. J Exp Zool 222:199–202. doi:10.1002/jez.1402220212
63. Sicard RE, Laffond WT (1983) Putative immunological influence upon amphibian forelimb regeneration. Pathobiology 51:337–344. doi:10.1159/000163213
64. Sicard RE, Lombard MF (1990) Putative immunological influence upon amphibian forelimb regeneration. II. Effects of X-irradiation on regeneration and allograft rejection. Biol Bull 178:21–24
65. Vinarsky V, Atkinson DL, Stevenson TJ et al (2005) Normal newt limb regeneration requires matrix metalloproteinase function. Dev Biol 279:86–98. doi:10.1016/j.ydbio.2004.12.003
66. Tsonis PA, Eguchi G (1982) Abnormal limb regeneration without tumor production in adult newts directed by carcinogens, 20-methylcholanthrene and benzo (α) pyrene. Develop Growth Differ 24:183–190. doi:10.1111/j.1440-169X.1982.00183.x
67. Pfeiffer CJ, Nagai T, Fujimura M, Tobe T (1985) Teratogenic effects of carcinogenic agents on limb regeneration in the japanese newt *Cynops pyrrhogaster*. Teratog Carcinog Mutagen 5:137–147. doi:10.1002/tcm.1770050303
68. Imokawa Y, Yoshizato K (1997) Expression of Sonic hedgehog gene in regenerating newt limb blastemas recapitulates that in developing limb buds. Proc Natl Acad Sci U S A 94:9159–9164
69. Lheureux E (1983) Replacement of irradiated epidermis by migration of non- irradiated epidermis in the newt limb: the necessity of healthy epidermis for regeneration. J Embryol Exp Morphol 76:217–223
70. Lheureux E, Thoms SD, Carey F (1986) The effects of two retinoids on limb regeneration in *Pleurodeles waltl* and *Triturus vulgaris*. J Embryol Exp Morphol 92:165–182
71. Zenjari C, Boilly B, Hondermarck H, Boilly Marer Y (1997) Nerve-blastema interactions induce fibroblast growth factor-1 release during limb regeneration in *Pleurodeles waltl*. Dev Growth Differ 39:15–22. doi:10.1046/j.1440-169X.1997.00003.x
72. Schotté OE, Butler EG (1941) Morphological effects of denervation and amputation of limbs in urodele larvae. J Exp Zool 87:279–322. doi:10.1002/jez.1400870207
73. Singer M (1943) The nervous system and regeneration of the forelimb of adult triturus. II. The role of the sensory supply. J Exp Zool 92:297–315. doi:10.1002/jez.1400920305
74. Bodemer CW (1958) The development of nerve-induced supernumerary limbs in the adult newt, *Triturus viridescens*. J Morphol 102:555–581. doi:10.1002/jmor.1051020304
75. Lebowitz P, Singer M (1970) Neurotrophic control of protein synthesis in the regenerating limb of the newt, *Triturus*. Nature 225:824–827
76. Flink IL (2002) Cell cycle reentry of ventricular and atrial cardiomyocytes and cells within the epicardium following amputation of the ventricular apex in the axolotl, *Ambystoma mexicanum*: confocal microscopic immunofluorescent image analysis of bromodeoxyuridine-labeled nuclei. Anat Embryol 205:235–244. doi:10.1007/s00429-002-0249-6
77. Cano-Martínez A, Vargas-González A (2010) Functional and structural regeneration in the axolotl heart (*Ambystoma mexicanum*) after partial ventricular amputation. Arch Cardiol Mex 80:79–86
78. Oberpriller JO, Oberpriller JC (1974) Response of the adult newt ventricle to injury. J Exp Zool 187:249–259. doi:10.1002/jez.1401870208
79. Bettencourt-Dias M, Mittnacht S, Brockes JP (2003) Heterogeneous proliferative potential in regenerative adult newt cardiomyocytes. J Cell Sci 116:4001–4009. doi:10.1242/jcs.00698
80. Witman N, Murtuza B, Davis B et al (2011) Recapitulation of developmental cardiogenesis governs the morphological and functional regeneration of adult newt hearts following injury. Dev Biol 354:67–76. doi:10.1016/j.ydbio.2011.03.021
81. Witman N, Heigwer J, Thaler B et al (2013) miR-128 regulates non-myocyte hyperplasia, deposition of extracellular matrix and Islet1 expression during newt cardiac regeneration. Dev Biol 383:253–263. doi:10.1016/j.ydbio.2013.09.011

82. Mercer SE, Odelberg SJ, Simon H-G (2013) A dynamic spatiotemporal extracellular matrix facilitates epicardial-mediated vertebrate heart regeneration. Dev Biol 382:457–469. doi:10.1016/j.ydbio.2013.08.002
83. Looso M, Preussner J, Sousounis K, Bruckskotten M, Michel CS, Lignelli E et al (2013) A de novo assembly of the newt transcriptome combined with proteomic validation identifies new protein families expressed during tissue regeneration. Genome Biol 14(2):R16
84. Looso M, Borchardt T, Krüger M, Braun T (2010) Advanced identification of proteins in uncharacterized proteomes by pulsed in vivo stable isotope labeling-based mass spectrometry. Mol Cell Proteomics 9:1157–1166. doi:10.1074/mcp.M900426-MCP200
85. Kirsche K, Kirsche W (1964) Regenerative processes in the telencephalon of *Ambystoma mexicanum*. J Hirnforsch 7:421–436
86. Maden M, Manwell LA, Ormerod BK (2013) Proliferation zones in the axolotl brain and regeneration of the telencephalon. Neural Dev 8:1. doi:10.1186/1749-8104-8-1
87. Parish CL, Beljajeva A, Arenas E, Simon A (2007) Midbrain dopaminergic neurogenesis and behavioural recovery in a salamander lesion-induced regeneration model. Development 134:2881–2887. doi:10.1242/dev.002329
88. Berg DA, Kirkham M, Beljajeva A et al (2010) Efficient regeneration by activation of neurogenesis in homeostatically quiescent regions of the adult vertebrate brain. Development 137:4127–4134. doi:10.1242/dev.055541
89. Berg DA, Kirkham M, Wang H et al (2011) Dopamine controls neurogenesis in the adult salamander midbrain in homeostasis and during regeneration of dopamine neurons. Cell Stem Cell 8:426–433. doi:10.1016/j.stem.2011.02.001
90. Hameed LS, Berg DA, Belnoue L et al (2015) Environmental changes in oxygen tension reveal ROS-dependent neurogenesis and regeneration in the adult newt brain. Elife. doi:10.7554/eLife.08422.001
91. Suetsugu-Maki R, Maki N, Nakamura K et al (2012) Lens regeneration in axolotl: new evidence of developmental plasticity. BMC Biol 10:1. doi:10.1186/1741-7007-10-103
92. Grogg MW, Call MK, Okamoto M et al (2005) BMP inhibition-driven regulation of six-3 underlies induction of newt lens regeneration. Nature 438:858–862. doi:10.1038/nature04175
93. Eguchi G, Eguchi Y, Nakamura K et al (2011) Regenerative capacity in newts is not altered by repeated regeneration and ageing. Nat Commun 2:384. doi:10.1038/ncomms1389
94. Sousounis K, Qi F, Yadav MC et al (2015) A robust transcriptional program in newts undergoing multiple events of lens regeneration throughout their lifespan. Elife. doi:10.7554/eLife.09594.001
95. Sousounis K, Michel CS, Bruckskotten M (2013) A microarray analysis of gene expression patterns during early phases of newt lens regeneration. Mol Vis 19:135–145
96. Madhavan M, Haynes TL, Frisch NC et al (2006) The role of Pax-6 in lens regeneration. Proc Natl Acad Sci U S A 103:14848–14853. doi:10.1073/pnas.0601949103
97. Nakamura K, Islam MR, Takayanagi M et al (2014) A transcriptome for the study of early processes of retinal regeneration in the adult newt, *Cynops pyrrhogaster*. PLoS One 9, e109831. doi:10.1371/journal.pone.0109831
98. Reyer RW (1948) An experimental study of lens regeneration in *Triturus viridescens viridescens*. I. Regeneration of a lens after lens extirpation in embryos and larvae of different ages. J Exp Zool 107:217–267. doi:10.1002/jez.1401070205
99. Reyer RW (1954) Further studies on lens development from the dorsal iris of *Triturus viridescens viridescens* in the absence of the embryonic lens. J Exp Zool 125:1–15. doi:10.1002/jez.1401250102
100. Eguchi G (1963) Electron microscopic studies on lens regeneration. Dev Growth Differ 8:45–62. doi:10.1111/j.1440-169X.1963.tb00185.x
101. Reyer RW (1966) The influence of neural retina and lens on lens regeneration from dorsal iris implants in *Triturus viridescens* larvae. Dev Biol 14:214–245. doi:10.1016/0012-1606(66)90014-5
102. Piatt J (1955) Regeneration of the spinal cord in the salamander. J Exp Zool 129:177–207. doi:10.1002/jez.1401290109

103. Echeverri K, Tanaka EM (2002) Ectoderm to mesoderm lineage switching during axolotl tail regeneration. Science 298:1993–1996. doi:10.1126/science.1077804
104. Mchedlishvili L, Epperlein HH, Telzerow A, Tanaka EM (2007) A clonal analysis of neural progenitors during axolotl spinal cord regeneration reveals evidence for both spatially restricted and multipotent progenitors. Development 134:2083–2093. doi:10.1242/dev.02852
105. Mchedlishvili L, Mazurov V, Grassme KS et al (2012) Reconstitution of the central and peripheral nervous system during salamander tail regeneration. Proc Natl Acad Sci U S A 109:E2258–E2266. doi:10.1073/pnas.1116738109/-/DCSupplemental/pnas.201116738SI.pdf
106. Rodrigo Albors A, Tazaki A, Rost F et al (2015) Planar cell polarity-mediated induction of neural stem cell expansion during axolotl spinal cord regeneration. Elife 4:664. doi:10.7554/eLife.10230
107. Sehm T, Sachse C, Frenzel C, Echeverri K (2009) miR-196 is an essential early-stage regulator of tail regeneration, upstream of key spinal cord patterning events. Dev Biol 334:468–480. doi:10.1016/j.ydbio.2009.08.008
108. Quiroz JFD, Tsai E, Coyle M et al (2014) Precise control of miR-125b levels is required to create a regeneration-permissive environment after spinal cord injury: a cross-species comparison between salamander and rat. Dis Model Mech 7:601–611. doi:10.1242/dmm.014837
109. Gearhart M, Erickson J, Walsh A, Echeverri K (2015) Identification of conserved and novel microRNAs during tail regeneration in the Mexican axolotl. Int J Mol Sci 2015(16):22046–22061. doi:10.3390/ijms160922046
110. Monaghan JR, Walker JA, Page RB et al (2007) Early gene expression during natural spinal cord regeneration in the salamander *Ambystoma mexicanum*. J Neurochem 101:27–40. doi:10.1111/j.1471-4159.2006.04344.x
111. Fei J-F, Schuez M, Tazaki A et al (2014) CRISPR-mediated genomic deletion of sox2 in the axolotl shows a requirement in spinal cord neural stem cell amplification during tail regeneration. Stem Cell Rep 3:444–459. doi:10.1016/j.stemcr.2014.06.018
112. Nicolas S, Papillon D, Perez Y et al (2003) The spatial restrictions of 5′HoxC genes expression are maintained in adult newt spinal cord. Biol Cell 95:389–394. doi:10.1016/j.biolcel.2003.09.004
113. Zukor KA, Kent DT, Odelberg SJ (2010) Fluorescent whole-mount method for visualizing three-dimensional relationships in intact and regenerating adult newt spinal cords. Dev Dyn 239:3048–3057. doi:10.1002/dvdy.22441
114. Zukor KA, Kent DT, Odelberg SJ (2011) Meningeal cells and glia establish a permissive environment for axon regeneration after spinal cord injury in newts. Neural Dev 6:1. doi:10.1186/1749-8104-6-1
115. Lepp AC, Carlone RL (2014) RARβ2 expression is induced by the down-regulation of microRNA 133a during caudal spinal cord regeneration in the adult newt. Dev Dyn 243:1581–1590. doi:10.1002/dvdy.24210
116. Chevallier S, Landry M, Nagy F, Cabelguen JM (2004) Recovery of bimodal locomotion in the spinal-transected salamander, *Pleurodeles waltlii*. Eur J Neurosci 20:1995–2007. doi:10.1111/j.1460-9568.2004.03671.x
117. Egar M, Singer M (1972) The role of ependyma in spinal cord regeneration in the urodele, *Triturus*. Exp Neurol 37:422–430. doi:10.1016/0014-4886(72)90085-4
118. Jordan HE, Beams HW (1930) Hepatectomy in the salamander with special reference to hemopoiesisandcytologyoftheliverremnant.ExpBiolMed28:181–184.doi:10.3181/00379727-28-5219
119. Williams DD (1961) Liver regeneration in the newt, *Triturus viridescens*. Physiol Zool 34:256–259. doi:10.2307/30152702
120. Goss RJ, Stagg MW (1958) Regeneration of lower jaws in adult newts. J Morphol 102:289–309. doi:10.1002/jmor.1051020204
121. Ghosh S, Thorogood P, Ferretti P (1994) Regenerative capability of upper and lower jaws in the newt. Int J Dev Biol 38:479–490
122. Ghosh S, Thorogood P, Ferretti P (1996) Regeneration of lower and upper jaws in urodeles is differentially affected by retinoic acid. Int J Dev Biol 40:1161–1170

123. Kurosaka H, Takano Yamamoto T, Yamashiro T, Agata K (2008) Comparison of molecular and cellular events during lower jaw regeneration of newt (*Cynops pyrrhogaster*) and West African clawed frog (*Xenopus tropicalis*). Dev Dyn 237:354–365. doi:10.1002/dvdy.21419
124. Finch RA (1969) The influence of the nerve on lower jaw regeneration in the adult newt, *Triturus viridescens*. J Morphol 129:401–413. doi:10.1002/jmor.1051290403
125. Lee J, Gardiner DM (2012) Regeneration of limb joints in the axolotl (*Ambystoma mexicanum*). PLoS One 7, e50615. doi:10.1371/journal.pone.0050615.g001
126. Hutchison C, Pilote M, Roy S (2007) The axolotl limb: a model for bone development, regeneration and fracture healing. Bone 40:45–56. doi:10.1016/j.bone.2006.07.005
127. Tsutsumi R, Inoue T, Yamada S, Agata K (2015) Reintegration of the regenerated and the remaining tissues during joint regeneration in the newt *Cynops pyrrhogaster*. Regeneration 2:26–36. doi:10.1002/reg2.28
128. Thornton CS (1938) The histogenesis of muscle in the regenerating fore limb of larval *Amblystoma punctatum*. J Morphol 62:17–47. doi:10.1002/jmor.1050620104
129. Carlson BM (1970) The regeneration of a limb muscle in the axolotl from minced fragments. Anat Rec 166:423–435. doi:10.1002/ar.1091660302
130. Diogo R, Nacu E, Tanaka EM (2014) Is salamander limb regeneration really perfect? Anatomical and morphogenetic analysis of forelimb muscle regeneration in GFP-transgenic axolotls as a basis for regenerative, developmental, and evolutionary studies. Anat Rec 297:1076–1089. doi:10.1002/ar.22906
131. Sandoval-Guzmán T, Wang H, Khattak S et al (2014) Fundamental differences in dedifferentiation and stem cell recruitment during skeletal muscle regeneration in two salamander species. Cell Stem Cell 14:174–187. doi:10.1016/j.stem.2013.11.007
132. Morrison JI, Lööf S, He P, Simon A (2006) Salamander limb regeneration involves the activation of a multipotent skeletal muscle satellite cell population. J Cell Biol 172:433–440. doi:10.1083/jcb.200509011
133. Morrison JI, Borg P, Simon A (2010) Plasticity and recovery of skeletal muscle satellite cells during limb regeneration. FASEB J 24:750–756. doi:10.1096/fj.09-134825
134. Calve S, Odelberg SJ, Simon H-G (2010) A transitional extracellular matrix instructs cell behavior during muscle regeneration. Dev Biol 344:259–271. doi:10.1016/j.ydbio.2010.05.007
135. Reyer RW (1954) Regeneration of the lens in the amphibian eye. Q Rev Biol 29:1–46. doi:10.2307/2812386
136. Goss RJ (1969) Principles of regeneration. Academic, New York
137. Todd TJ (1823) On the process of reproduction of the members of the aquatic salamander. Q J Sci Lit Arts 16:84–96
138. Singer M (1952) The influence of the nerve in regeneration of the amphibian extremity. Q Rev Biol 27:169–200. doi:10.2307/2811581
139. Singer M, Rzehak K, Maier CS (1967) The relation between the caliber of the axon and the trophic activity of nerves in limb regeneration. J Exp Zool 166:89–97. doi:10.1002/jez.1401660110
140. Brockes JP (1984) Mitogenic growth factors and nerve dependence of limb regeneration. Science 225:1280–1287
141. Tassava RA, Mccullough WD (1978) Neural control of cell cycle events in regenerating salamander limbs. Am Zool 18:843–854. doi:10.1093/icb/18.4.843
142. Locatelli DP (1929) Der Einfluss des nervensystems auf die regeneration. Wilhelm Roux Arch Entwickl Mech Org 114:686–770. doi:10.1007/BF02078924
143. Bodemer CW (1959) Observations on the mechanism of induction of supernumerary limbs in adult *Triturus viridescens*. J Exp Zool 140:79–99. doi:10.1002/jez.1401400105
144. Cohen N (1969) Immunogenetic and developmental aspects of tissue transplantation immunity in urodele amphibians. In: Mizell M (ed) Biology of amphibian tumors. Springer, Berlin Heidelberg, pp 153–168.
145. Kinefuchi K, Kushida Y, Touma M, Hosono M (2013) Limited immune diversity in Urodela: chronic transplantation responses occur even with family-disparate xenografts. Zoolog Sci 30:577–584. doi:10.2108/zsj.30.577
146. Butler EG (1951) The mechanics of blastema formation and regeneration in urodele limbs of reversed polarity. Trans NY Acad Sci 13:164–167. doi:10.1111/j.2164-0947.1951.tb01015.x

147. Bohn H (1972) The origin of the epidermis in the supernumerary regenerates of triple legs in cockroaches (*Blattaria*). J Embryol Exp Morphol 28:185–208
148. Lheureux E (1972) Contribution a l'etude du role de la peau et des tissus axiaux du membre dans le declenchement de morphogeneses regeneratrices anormales chez le triton *Pleurodeles waltii*. Ann Embryol Morphog 5:165–172
149. Niazi IA, Saxena S (1978) Abnormal hind limb regeneration in tadpoles of the toad, *Bufo andersoni*, exposed to excess vitamin A. Folia Biol (Krakow) 26:3–8
150. Kim W-S, Stocum DL (1986) Retinoic acid modifies positional memory in the anteroposterior axis of regenerating axolotl limbs. Dev Biol 114:170–179. doi:10.1016/0012-1606(86)90393-3
151. Tsonis PA, Trombley MT, Rowland T et al (2000) Role of retinoic acid in lens regeneration. Dev Dyn 219:588–593. doi:10.1002/1097-0177(2000)9999:9999<::AID-DVDY1082>3.0.CO;2-H
152. Garza-Garcia A, Harris R, Esposito D et al (2009) Solution structure and phylogenetics of Prod1, a member of the three-finger protein superfamily implicated in salamander limb regeneration. PLoS One 4, e7123. doi:10.1371/journal.pone.0007123.s010
153. Geng J, Gates PB, Kumar A et al (2015) Identification of the orphan gene Prod 1 in basal and other salamander families. Evodevo 6:9. doi:10.1186/s13227-015-0006-6
154. Sobkow L, Epperlein HH, Herklotz S et al (2006) A germline GFP transgenic axolotl and its use to track cell fate: dual origin of the fin mesenchyme during development and the fate of blood cells during regeneration. Dev Biol 290:386–397. doi:10.1016/j.ydbio.2005.11.037
155. Sicard RE (1983) Blood cells and their role in regeneration. Pathobiology 51:51–59. doi:10.1159/000163173
156. Debuque RJ, Godwin JW (2015) Methods for axolotl blood collection, intravenous injection, and efficient leukocyte isolation from peripheral blood and the regenerating limb. In: Kumar A, Simon A (eds) Salamanders in regeneration research. Springer, New York, NY, pp 205–226
157. Lopez D, Lin L, Monaghan JR et al (2014) Mapping hematopoiesis in a fully regenerative vertebrate: the axolotl. Blood 124:1232–1241. doi:10.1182/blood-2013-09-526970
158. Diaz Quiroz JF, Tsai E, Coyle M et al (2014) Precise control of miR-125b levels is required to create a regeneration-permissive environment after spinal cord injury: a cross-species comparison between salamander and rat. Dis Model Mech 7:601–611. doi:10.1242/dmm.014837
159. Henry JJ, Tsonis PA (2010) Molecular and cellular aspects of amphibian lens regeneration. Prog Retin Eye Res 29:543–555. doi:10.1016/j.preteyeres.2010.07.002
160. Licht LE, Lowcock LA (1991) Genome size and metabolic rate in salamanders. Comp Biochem Physiol B 100:83–92. doi:10.1016/0305-0491(91)90089-V
161. Sun C, Shepard DB, Chong RA et al (2012) LTR retrotransposons contribute to genomic gigantism in plethodontid salamanders. Genome Biol Evol 4:168–183. doi:10.1093/gbe/evr139
162. Smith JJ, Putta S, Zhu W et al (2009) Genic regions of a large salamander genome contain long introns and novel genes. BMC Genomics 10:19. doi:10.1186/1471-2164-10-19
163. Zhang P, Wake DB (2009) Higher-level salamander relationships and divergence dates inferred from complete mitochondrial genomes. Mol Phylogenet Evol 53:492–508. doi:10.1016/j.ympev.2009.07.010
164. Abdullayev I, Kirkham M, Björklund ÅK et al (2013) A reference transcriptome and inferred proteome for the salamander *Notophthalmus viridescens*. Exp Cell Res 319:1187–1197. doi:10.1016/j.yexcr.2013.02.013
165. Baddar NWAH, Woodcock MR, Khatri S et al (2015) Sal-site: research resources for the Mexican axolotl. In: Kumar A, Simon A (eds) Salamanders in regeneration research. Springer, New York, NY, pp 321–336
166. Bruckskotten M, Looso M, Reinhardt R et al (2012) Newt-omics: a comprehensive repository for omics data from the newt *Notophthalmus viridescens*. Nucl Acids Res 40:D895–D900. doi:10.1093/nar/gkr873
167. Looso M, Braun T (2015) Data mining in newt-omics, the repository for omics data from the newt. In: Kumar A, Simon A (eds) Salamanders in regeneration research. Springer, New York, NY, pp 337–351
168. Rao N, Song F, Jhamb D et al (2014) Proteomic analysis of fibroblastema formation in regenerating hind limbs of *Xenopus laevis* froglets and comparison to axolotl. BMC Dev Biol 14:32. doi:10.1186/1471-213X-14-32

169. Looso M, Michel CS, Konzer A et al (2012) Spiked-in pulsed in vivo labeling identifies a new member of the ccn family in regenerating newt hearts. J Proteome Res 11:4693–4704. doi:10.1021/pr300521p
170. Keinath MC, Timoshevskiy VA, Timoshevskaya NY et al (2015) Initial characterization of the large genome of the salamander *Ambystoma mexicanum* using shotgun and laser capture chromosome sequencing. Sci Rep 5:16413. doi:10.1038/srep16413
171. Roy S, Gardiner DM, Bryant SV (2000) Vaccinia as a tool for functional analysis in regenerating limbs: ectopic expression of Shh. Dev Biol 218:199–205. doi:10.1006/dbio.1999.9556
172. Laube F, Heister M, Scholz C et al (2006) Re-programming of newt cardiomyocytes is induced by tissue regeneration. J Cell Sci 119:4719–4729. doi:10.1242/jcs.03252
173. Whited JL, Tsai SL, Beier KT et al (2013) Pseudotyped retroviruses for infecting axolotl in vivo and in vitro. Development 140:1137–1146. doi:10.1242/dev.087734
174. Khattak S, Sandoval-Guzmán T, Stanke N et al (2013) Foamy virus for efficient gene transfer in regeneration studies. BMC Dev Biol 13:17. doi:10.1128/JVI.76.8.3774-3783.2002
175. Casco Robles MM, Yamada S, Miura T, Chiba C (2010) Simple and efficient transgenesis with I-SceI meganuclease in the newt, *Cynops pyrrhogaster*. Dev Dyn 239:3275–3284. doi:10.1002/dvdy.22463
176. Hayashi T, Yokotani N, Tane S et al (2013) Molecular genetic system for regenerative studies using newts. Dev Growth Differ 55:229–236. doi:10.1111/dgd.12019
177. Hayashi T, Sakamoto K, Sakuma T et al (2014) Transcription activator-like effector nucleases efficiently disrupt the target gene in Iberian ribbed newts (*Pleurodeles waltl*), an experimental model animal for regeneration. Dev Growth Differ 56:115–121. doi:10.1111/dgd.12103
178. Flowers GP, Timberlake AT, McLean KC et al (2014) Highly efficient targeted mutagenesis in axolotl using Cas9 RNA-guided nuclease. Development 141:2165–2171. doi:10.1242/dev.105072
179. Khattak S, Murawala P, Andreas H et al (2014) Optimized axolotl (*Ambystoma mexicanum*) husbandry, breeding, metamorphosis, transgenesis and tamoxifen- mediated recombination. Nat Protoc 9:529–540. doi:10.1038/nprot.2014.040
180. Schnapp E, Tamaka EM (2004) Quantitative evaluation of morpholino-mediated protein knockdown of GFP, MSX1, and PAX7 during tail regeneration in *Ambystoma mexicanum*. Dev Dyn 232:162–170. doi:10.1002/dvdy.20203
181. Lam NT, Currie PD, Lieschke GJ et al (2012) Nerve growth factor stimulates cardiac regeneration via cardiomyocyte proliferation in experimental heart failure. PLoS One 7, e53210. doi:10.1371/journal.pone.0053210
182. Mahmoud AI, O'Meara CC, Gemberling M et al (2015) Nerves regulate cardiomyocyte proliferation and heart regeneration. Dev Cell 34:387–399. doi:10.1016/j.devcel.2015.06.017
183. Takeo M, Chou WC, Sun Q et al (2013) Wnt activation in nail epithelium couples nail growth to digit regeneration. Nature 499:228–232. doi:10.1038/nature12214
184. Buckley G, Wong J, Metcalfe AD, Ferguson MWJ (2012) Denervation affects regenerative responses in MRL/MpJ and repair in C57BL/6 ear wounds. J Anat 220:3–12. doi:10.1111/j.1469-7580.2011.01452.x
185. Yamazaki S, Ema H, Karlsson G et al (2011) Nonmyelinating schwann cells maintain hematopoietic stem cell hibernation in the bone marrow niche. Cell 147:1146–1158. doi:10.1016/j.cell.2011.09.053
186. Brownell I, Guevara E, Bai CB et al (2011) Nerve-derived sonic hedgehog defines a niche for hair follicle stem cells capable of becoming epidermal stem cells. Cell Stem Cell 8:552–565. doi:10.1016/j.stem.2011.02.021
187. Mescher AL, Neff AW, King MW (2013) Changes in the inflammatory response to injury and its resolution during the loss of regenerative capacity in developing *Xenopus* limbs. PLoS One 8, e80477. doi:10.1371/journal.pone.0080477.t002
188. Love NR, Chen Y, Ishibashi S et al (2013) Amputation-induced reactive oxygen species are required for successful *Xenopus* tadpole tail regeneration. Nat Cell Biol 15:222–228. doi:10.1038/ncb2659

189. Petrie TA, Strand NS, Tsung-Yang C et al (2014) Macrophages modulate adult zebrafish tail fin regeneration. Development 141:2581–2591. doi:10.1242/dev.098459
190. Heredia JE, Mukundan L, Chen FM et al (2013) Type 2 innate signals stimulate fibro/adipogenic progenitors to facilitate muscle regeneration. Cell 153:376–388. doi:10.1016/j.cell.2013.02.053
191. Aurora AB, Porrello ER, Tan W et al (2014) Macrophages are required for neonatal heart regeneration. J Clin Invest 124:1382–1392. doi:10.1172/JCI72181

Chapter 2
Regeneration: Lessons from the Lizard

Elizabeth D. Hutchins, Jeanne Wilson-Rawls, and Kenro Kusumi

2.1 Regeneration in Lizards

Regeneration of entire appendages requires complex coordination of molecular events including activation of stem cells or dedifferentiation to form proliferative cells, proliferation, and differentiation into the musculoskeletal, nervous, and epithelial tissues of the regenerated structure. The ability to regenerate entire appendages is a common trait found in teleost fish, amphibians, and squamate reptiles [1, 2]. The ability to regenerate an appendage can vary between different periods of its lifespan and between anatomical structures. These vertebrates have a common ancestor and their shared evolutionary history is reflected in their genomes, sharing multiple homologous genetic pathways that regulate developmental patterning and differentiation [3].

In the past decade, appendage regeneration research in reptiles has focused on describing tail regeneration in lizards using the green anole, *Anolis carolinensis*, (Fig. 2.1; [4–7]) and the leopard gecko, *Eublepharis macularius*, [8–11] as models. The green anole is used as model of development [12, 13], population genetics [14, 15], reproductive physiology and behavior [16, 17], and functional morphology [7, 18], and it was the first non-avian reptile to have a sequenced genome [19]. The availability of the genome makes molecular genetic studies of regeneration feasible. There have been a broad scope of studies of the green anole [20–32] that inform

E.D. Hutchins • K. Kusumi (✉)
School of Life Sciences, Arizona State University, Tempe, AZ 85287, USA

Neurogenomics Division, Translational Genomics Research Institute,
Phoenix, AZ 85004, USA
e-mail: ehutchins@tgen.org; kenro@asu.edu

J. Wilson-Rawls
School of Life Sciences, Arizona State University, Tempe, AZ 85287, USA
e-mail: jeanne.wilson-rawls@asu.edu

J. Wilson-Rawls, K. Kusumi (eds.), *Innovations in Molecular Mechanisms and Tissue Engineering*, Stem Cell Biology and Regenerative Medicine,
DOI 10.1007/978-3-319-44996-8_2

Fig. 2.1 Image of a green anole lizard with a regenerated tail. The *arrow* indicates the autotomy break point and start of the regenerated tail. Photo credit: Joel Robertson

more recent molecular, cellular, and anatomical analyses [5–7]. In alligators while tail regeneration has been reported, the structure and process of regeneration are unknown [33, 34]. The regenerated lizard tail is an extraordinary example of de novo development of hyaline/articular cartilage, muscle groups with tendinous attachments, skin, vasculature, and neural ependymal cells [5, 7, 9, 11]. In contrast, birds and mammals have very limited regenerative capacity. Regeneration in mammals is restricted to neonatal and juvenile individuals, including the regrowth of digit tips [35–38].

2.2 Stages of Regeneration in Lizards

Lizards represent the evolutionarily closest related group to mammals that demonstrate the ability to regenerate appendages (Fig. 2.1). Many lizard species can undergo tail autotomy followed by regeneration [33]; this is a self-induced amputation induced by physiological and/or mechanical stress leading to shedding of the tail as a predator evasion tactic. The vertebrae in the tail of many lizards have fracture planes that permit autotomy [30]. Following autotomy, there is a well described process of tail regeneration that displays aspects of a two step model of regeneration (Fig. 2.2; reviewed in [39]). In this model, there is an initial immune response following injury leading to either scar formation or full regeneration. The regenerative response includes (1) capping of the wound with a blood clot and remodeling of the ECM, (2) emergence of a wound epithelium and loss of the scab, (3) generation of proliferating cells, blood vessel formation, and thickening of the wound epithelium, and (4) growth and differentiation of tissues in the growing tail, including the neuroependyma, cartilage, and myofibers [11, 40, 41]. Studies in the leopard gecko demonstrate that tail regeneration is not limited to loss at autotomy planes; regeneration will occur whether or not the loss occurs close to the fracture plane, the tail is amputated mechanically, or it is released via autotomy [8]. In contrast to tail autotomy, the amputation of the limb leads to initial injury responses with partial formation of some tissues but ending in scar formation in the viviparous lizard *Lacerta vivipara* [42] and the common wall lizard *Podarcis muralis* [43].

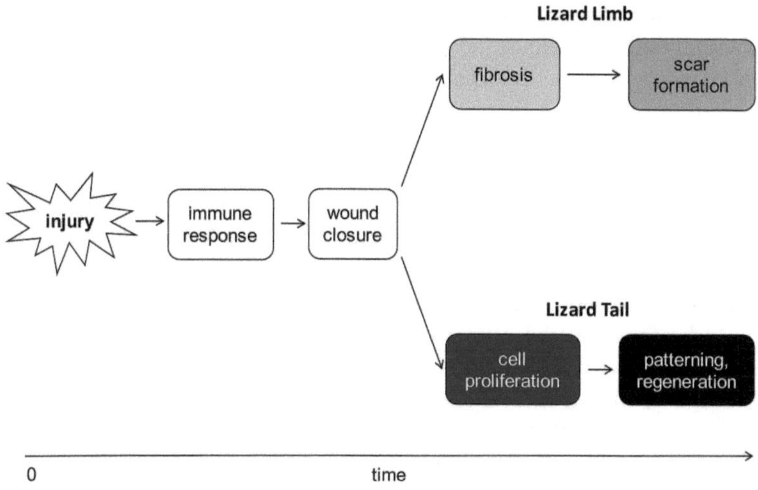

Fig. 2.2 Two step model of regeneration. Lizards are able to regenerate their tails following autotomy. However, following limb amputation, lizards display an injury response with partial regrowth but followed by fibrosis and scarring. In contrast, tail autotomy is followed by formation of the wound epithelium, ECM remodeling, then cell proliferation and patterning in regeneration

2.3 Molecular Mechanisms of Lizard Regeneration

Prior to the outgrowth observed in regeneration, the damaged tissue is covered by a wound epithelium for scar-free wound healing [44, 45]. This wound epithelium expands in thickness to twice that of the original epidermis in the lizard and newt [8, 11, 40]. In the newt, this structure has been called the apical epithelial cap (AEC), in reference to the apical ectodermal ridge (AER) formed at the edge of the limb bud development [46–48].

Remodeling and clean-up of the damaged tissues takes place before the onset of outgrowth in regeneration. Key to this process is the reorganization of extracellular matrix (ECM) to create a new scaffolding matrix for the regenerated appendage [49, 50]. Remodeling of the ECM is a characteristic of the scar-free wound healing that occurs prior to regeneration, as opposed to a fibrotic, non-regenerative response [51, 52]. Several factors regulating scar-free wound healing have been identified. Matrix metalloproteases (MMPs), which have been observed in the regenerating tail of the green anole lizard [6] and the leopard gecko [8], likely contribute to ECM remodeling. In addition to ECM remodeling, regulation of the inflammatory response and inhibition of fibrosis are key early steps that permit scar-free regeneration [53–58]. Studies in the Italian wall lizard (*Podarcis sicula*) have identified infiltration of granulocytes and monocytes/macrophages into the autotomized tail stump [59, 60]. Given their role in regulation of inflammation, ECM remodeling, fibroblast formation, angiogenesis, and peripheral nerve innervation, macrophages are of particular interest [61–64]. Macrophages regulate proliferation of endothelial cells, keratinocytes, and fibroblasts [65] as well as stimulate the production of immune cytokines including PDGFs, IGFs, FGFs, TGFs, CSFs, hepatocyte growth factors, colony-stimulating factors, and Wnt ligands [66].

The formation of the blastema is well described in the regenerating limbs and fins of amphibians and teleost fish. A blastema is traditionally defined as the dedifferentiated coalescence of pluripotent proliferating cells concentrated at the tip of a regenerating appendage. Importantly, there is a lack of a vascular bed found at the distal tip [67–77]. More recently, studies in amphibians have found that the traditional view of the blastema as a mass of pluripotential, dedifferentiated cells is not entirely accurate. Further, the cellular composition of this structure can vary by stage and species [78, 79]. For example, in the newt, *Notophthalmus viridescens*, studies with Cre/loxP mediated lineage tracing found that mature muscle of an amputated limb dedifferentiated and formed PAX7-negative proliferating cells that could be found in the blastema. However, these cells contributed solely to regenerating muscle [80]. Whereas, in the axolotl, *Ambystoma mexicanum,* these same lineage tracing approaches demonstrated that the remaining muscle did not dedifferentiate, nor contribute any cells to the blastema. Muscle regeneration in this salamander occurs through PAX7-positive satellite cells, the resident stem cell population found in muscle [80]. This was also observed when transplanted GFP-positive cells were used to track cells in regenerating axolotl limbs. These studies demonstrated that all cells that contributed to the blastema retained their original embryological fate and contributed only to those tissues. Cells that were derived from lateral plate mesoderm only contributed to dermis, and skeleton and muscle precursors that are derived from presomitic mesoderm only became muscle [78]. Interestingly, in the Japanese newt, *Cynops pyrrhogaster*, post-metamorphosis muscle regeneration in amputated limbs occurs through muscle dedifferentiation, but pre-metamorphosis PAX7-positive satellite cells regenerate muscle post-amputation [79].

Clearly de-differentiation as a source of proliferating progenitor cells is not the rule, and this is consistent with observations from studies of *A. carolinensis* tail regeneration [81–84]. In histological sections it was noted that differentiating muscle was apparent as early as 15 days post autotomy (dpa); regenerating tails in this species demonstrate significant distal outgrowth until 65 dpa [5]. By 20 dpa, there was differentiating muscle from the distal tip to the proximal breakpoint, but there was no obvious zone of proliferating progenitors at the tip [6]. Interestingly, the distal tip of the regenerating tail is also highly vascularized (Fig. 2.3) [6]. Cartilage, which replaces the missing skeleton, and the ependymal cells that regenerate the spinal cord, extend from the breakpoint to the distal tip of the early regenerating tail (20 dpa) [6]. Proliferating cells were found throughout the regenerating anole tail when assayed using an antibody that recognized MCM2, a protein expressed in cells that are replicating their genome in preparation to divide. Subsequent transcriptome analysis of genes involved in proliferation complemented these data; it was found that these genes were expressed at similar levels all along the tail. Interestingly, the lowest level of expression was found in the region of the distal tip [6]. Similarly in the leopard gecko, proliferating cells were found throughout the regenerating tail, instead of restricted to the distal tip, and the distal tip is vascularized as well [11]. In these lizards a true blastema does not seem to exist.

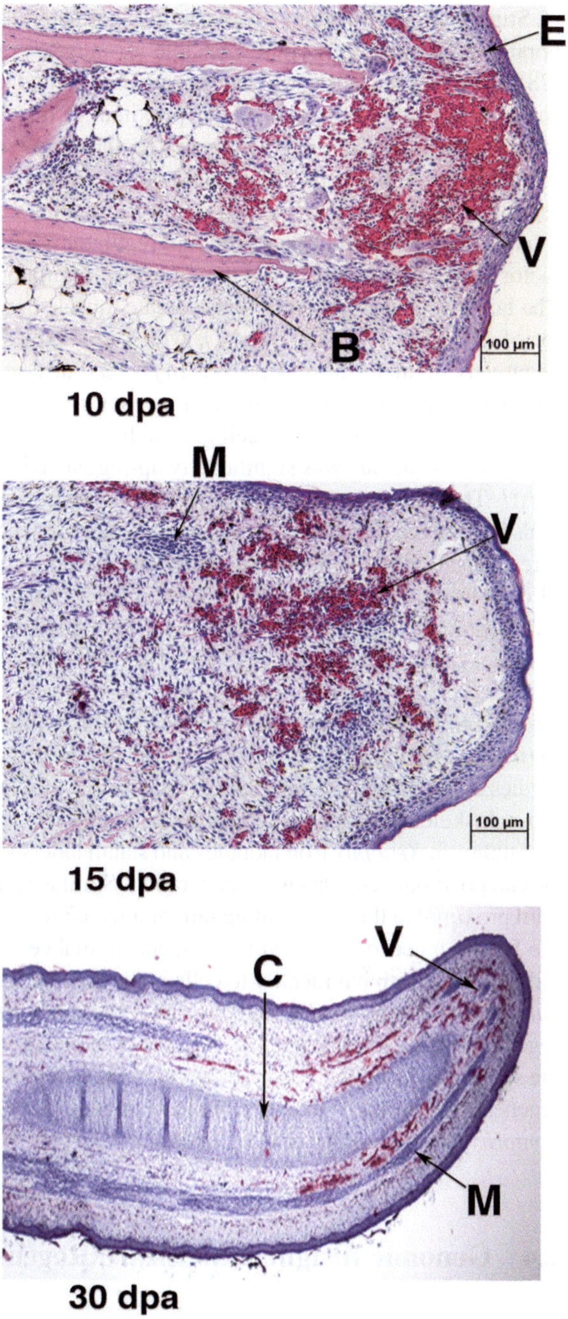

Fig. 2.3 Histology of early regenerating *A. carolinensis* tail. Sagittal sections through the tips of the early regenerating tail stained with H&E. At 10 dpa the outgrowth of the tail has not started but the epithelium (E) has regenerated and the distalmost portion is highly vascularized (V). At 15 dpa outgrowth has started, there are muscle (M) groups developing near the distal tip and the vascular network is still prominent. At 30 dpa, there is well developed muscle and cartilage (C). The vascular network has extended

Studies in salamanders demonstrated that satellite cells, an existing progenitor population in muscle, were responsible for muscle regeneration in amputated limbs [78]. Based on our observations, regeneration in anole lizard tails employs a similar strategy. Transcriptome analysis of proximal to distal gene expression in the early regenerating tail (25 dpa) demonstrated that there was significant expression of markers of satellite cells and muscle development. These genes include important regulatory factors such as the marker of mammalian satellite cells paired box domain 7 (*pax7*), the myogenic transcriptional regulator MyoD (*myod1*), myocyte enhancer factor 2C (*mef2c)* a cofactor of the myogenic regulators, twist1, and Mohawk (*mkx*). The tail also expresses genes that regulate muscle development such as nuclear factor of activated T cells 1 (*nfatc1)*, which regulates skeletal muscle fiber type and negatively regulates MyoD, paraxis (*tcf15*) a transcription factor that regulates compartmentalization of the somite, and myostatin (*mstn*), a TGFβ family member and negative regulator of muscle cell growth [6].

Another gene that was significantly up-regulated in the regenerating anole tail was *twist1*. This gene encodes a basic helix-loop-helix transcription factor that in mammals is involved in limb patterning and Saethre-Chozen syndrome [85–90]. There are three Twist family members and *Twist1* and *Twist3* were found in specific populations of cells in the ambystoma limb blastema [91]. Using single cell PCR, it was found that blastemal cells that expressed *Twist1* and *Sox9* and were derived from, and will become, cartilage whereas *Myf5* positive cells that will become muscle did not co-express *Twist1 or Twist3*. Consistently, *Twist1* and *Twist3* co-expressing cells were destined to become dermis and were derived from this tissue [91]. In the anole tail, *twist1* was significantly up-regulated in the regenerating tail [6], a challenge for future studies in the lizard will be to identify the source of stem/progenitor cells for different musculoskeletal tissues in the regenerating tail.

Studies in *Xenopus* frog tadpoles and salamanders suggest that nerve signaling is a crucial positional cue driving regeneration. Similarly, in lizards, damage to the spinal cord proximal to the regenerating tail inhibits the regenerative process [28, 92, 93]. In the Japanese gecko, *Gekko japonicus*, ependymal cells at the core of the regenerating tail provide positional identity to cells in the regenerating tail [94]. Studies done in *A. carolinensis* and *Scincella lateralis* have shown that the ependyma is necessary for regeneration of the cartilage [28, 81, 84]. The ependymal cells regrow directly from the spinal cord, and there is no evidence of dedifferentiation of nervous tissues in tail regeneration in many lizards examined including *A. carolinensis*, *Sphaerodactylus goniorhynchus, S. argus* and *Lygosoma laterale* [30, 81, 82].

2.4 Genomic Insights into Lizard Regeneration

With the availability of high throughput sequencing technologies and emergence of annotated genomes for regenerative species, gene expression studies of regeneration in reptiles have become possible [95]. In the green anole lizard,

RNA-Seq analysis has identified at least 326 genes that are differentially expressed within different regions of the regenerating tail, including regulators of muscle and cartilage development, wound response, and thyroid hormonal response, and members of the Wnt and FGF/MAPK pathways. These data can be compared to similar gene expression studies in other regenerative model organisms in order to identify common factors required for regeneration across vertebrates. Namely, studies in a number of vertebrate models have identified genes in the Wnt-Ca^{2+} pathway in both regeneration and regulation of the inflammatory response [96–98]. In the green anole lizard, *wnt5a* and the Wnt inhibitors *dkk2 and cerberus* were elevated in the distal tip of the regenerating tail [6]. *Wnt5a* and *wnt5b* are expressed in the axolotl limb blastema [99], and in the regenerating fins of zebrafish *wnt5a*, *wnt5b*, and *wnt10* are co-expressed [100]. Further studies will help to identify the role that Wnt signaling plays in creating permissible conditions for regeneration.

Given the large number of genes differentially expressed during the process of regeneration, attention has been focused on regulatory agents such as microRNAs, which are highly conserved amongst metazoans and can modulate the expression of multiple genes [101]. MicroRNAs have been found to regulate a number of biological processes, including proliferation and differentiation in cells ranging from skeletal and cardiac muscle to neurons [102], hematopoietic and embryonic stem cells [103, 104] and T cells [105], as well as repair of muscle [106]. MicroRNAs has also been found in regeneration of the limb and tail of axolotl salamanders [107, 108], lens and inner ear of newts [109, 110], and tail, spinal cord, and heart in the zebrafish [111–113]. Recently, sequencing in the green anole lizard regenerating tail and adult tissues has identified 350 putative novel and 196 known microRNAs [114]. In the regenerating tail at peak growth (25 days post autotomy), 11 differentially expressed microRNAs were identified within the growing tail, including miR-133a, miR-133b, and miR-206, a regulator of stem cell proliferation in other regenerating species. In addition, 3 novel microRNAs were identified to be elevated in the tail tip, suggesting potentially uncharacterized pathways or regulators specific to lizards may involved in regeneration.

MicroRNAs are not the only factors that may lead to differential expression of hundreds of genes; lizards and other regenerative species could potentially display genomic changes in coding or non-coding regulatory sequences such as enhancers, silencers, and insulators that account for regenerative differences. Alternately, changes in chromatin regulation between regenerative and non-regenerative vertebrates may also play a role. Further comparative studies making use of multiple model systems will allow us to distinguish between these possibilities.

Acknowledgments We thank Joel Robertson for his photograph of the green anole.

References

1. Agata K, Inoue T (2012) Survey of the differences between regenerative and non-regenerative animals. Dev Growth Differ 54(2):143–152. doi:10.1111/j.1440-169X.2011.01323.x
2. Bely AE, Nyberg KG (2010) Evolution of animal regeneration: re-emergence of a field. Trends Ecol Evol 25(3):161–170. doi:10.1016/j.tree.2009.08.005
3. Cañestro C, Yokoi H, Postlethwait JH (2007) Evolutionary developmental biology and genomics. Nat Rev Genet 8:932–942. doi:10.1038/nrg2226
4. Eckalbar WL, Hutchins ED, Markov GJ, Allen AN, Corneveaux JJ, Lindblad-Toh K, Di Palma F, Alföldi J, Huentelman MJ, Kusumi K (2013) Genome reannotation of the lizard *Anolis carolinensis* based on 14 adult and embryonic deep transcriptomes. BMC Genomics 14:49. doi:10.1186/1471-2164-14-49
5. Fisher RE, Geiger LA, Stroik LK, Hutchins ED, George RM, Denardo DF, Kusumi K, Rawls JA, Wilson-Rawls J (2012) A histological comparison of the original and regenerated tail in the green anole, *Anolis carolinensis*. Anat Rec (Hoboken) 295(10):1609–1619. doi:10.1002/ar.22537
6. Hutchins ED, Markov GJ, Eckalbar WL, George RM, King JM, Tokuyama MA, Geiger LA, Emmert N, Ammar MJ, Allen AN, Siniard AL, Corneveaux JJ, Fisher RE, Wade J, DeNardo DF, Rawls JA, Huentelman MJ, Wilson-Rawls J, Kusumi K (2014) Transcriptomic analysis of tail regeneration in the lizard *Anolis carolinensis* reveals activation of conserved vertebrate developmental and repair mechanisms. PLoS One 9, e105004. doi:10.1371/journal.pone.0105004.s014
7. Ritzman TB, Stroik LK, Julik E, Hutchins ED, Lasku E, Denardo DF, Wilson-Rawls J, Rawls JA, Kusumi K, Fisher RE (2012) The gross anatomy of the original and regenerated tail in the green anole (*Anolis carolinensis*). Anat Rec (Hoboken) 295(10):1596–1608. doi:10.1002/ar.22524
8. Delorme SL, Lungu IM, Vickaryous MK (2012) Scar-free wound healing and regeneration following tail loss in the leopard gecko, *Eublepharis macularius*. Anat Rec (Hoboken) 295(10):1575–1595. doi:10.1002/ar.22490
9. Gilbert EAB, Payne SL, Vickaryous MK (2013a) The anatomy and histology of caudal autotomy and regeneration in lizards. Physiol Biochem Zool 86(6):631–644. doi:10.1086/673889
10. Gilbert RWD, Vickaryous MK, Viloria-Petit AM (2013b) Characterization of TGFβ signaling during tail regeneration in the leopard gecko (*Eublepharis macularius*). Dev Dyn 242(7):886–896. doi:10.1002/dvdy.23977
11. McLean KE, Vickaryous MK (2011) A novel amniote model of epimorphic regeneration: the leopard gecko, *Eublepharis macularius*. BMC Dev Biol 11(1):50. doi:10.1186/1471-213X-11-50
12. Eckalbar WL, Lasku E, Infante CR, Elsey RM, Markov GJ, Allen AN, Corneveaux JJ, Losos JB, DeNardo DF, Huentelman MJ, Wilson-Rawls J, Rawls A, Kusumi K (2012) Somitogenesis in the anole lizard and alligator reveals evolutionary convergence and divergence in the amniote segmentation clock. Dev Biol 363(1):308–319. doi:10.1016/j.ydbio.2011.11.021
13. Sanger TJ, Losos JB, Gibson-Brown JJ (2008) A developmental staging series for the lizard genusAnolis: a new system for the integration of evolution, development, and ecology. J Morphol 269:129–137. doi:10.1002/jmor.10563
14. Tollis M, Boissinot S (2014) Genetic variation in the green anole lizard (*Anolis carolinensis*) reveals island refugia and a fragmented Florida during the quaternary. Genetica 142(1):59–72. doi:10.1007/s10709-013-9754-1
15. Wordley C, Slate J, Stapley J (2011) Mining online genomic resources in *Anolis carolinensis* facilitates rapid and inexpensive development of cross-species microsatellite markers for the Anolis lizard genus. Mol Ecol Resour 11(1):126–133. doi:10.1111/j.1755-0998.2010.02863.x
16. Lovern MB, Wade J (2003) Yolk testosterone varies with sex in eggs of the lizard, *Anolis carolinensis*. J Exp Zool A Comp Exp Biol 295(2):206–210. doi:10.1002/jez.a.10225
17. Wade J (2012) Sculpting reproductive circuits: relationships among hormones, morphology and behavior in anole lizards. Gen Comp Endocrinol 176(3):456–460. doi:10.1016/j.ygcen.2011.12.011

18. Montuelle SJ, Herrel A, Libourel PA, Reveret L, Bels VL (2009) Locomotor-feeding coupling during prey capture in a lizard (*Gerrhosaurus major*): effects of prehension mode. J Exp Biol 212(Pt 6):768–777. doi:10.1242/jeb.026617
19. Alföldi J, Di Palma F, Grabherr M et al (2011) The genome of the green anole lizard and a comparative analysis with birds and mammals. Nature 477(7366):587–591. doi:10.1038/nature10390
20. Alibardi L (1995) Development of the axial cartilaginous skeleton in the regenerating tail of lizards. Bull Assoc Anat (Nancy) 79(244):3–9
21. Alibardi L (1995) Muscle differentiation and morphogenesis in the regenerating tail of lizards. J Anat 186(Pt 1):143–151
22. Alibardi L (2010) Morphological and cellular aspects of tail and limb regeneration in lizards. A model system with implications for tissue regeneration in mammals. Adv Anat Embryol Cell Biol 207:iii–v–x– 1–109
23. Alibardi L (2014) Observations on lumbar spinal cord recovery after lesion in lizards indicates regeneration of a cellular and fibrous bridge reconnecting the injured cord. J Dev Biol 2(4):210–229. doi:10.3390/jdb2040210
24. Alibardi L (2014) Ultrastructural observations on lumbar spinal cord recovery after lesion in lizard indicates axonal regeneration and neurogenesis. Int J Biol 7(1). doi:10.5539/ijb.v7n1p122
25. Chlebowski JS, Przbylski RJ, Cox PG (1973) Ultrastructural studies of lizard (I) myogenesis in vitro. Dev Biol 33(1):80–99
26. Cox PG (1968) In vitro myogenesis of promuscle cells from the regenerating tail of the lizard, Anolis carolinensis. J Morphol 126(1):1–18. doi:10.1002/jmor.1051260102
27. Egar M, Simpson SB Jr, Singer M (1970) The growth and differentiation of the regenerating spinal cord of the lizard, *Anolis carolinensis*. J Morphol 131(2):131–151. doi:10.1002/jmor.1051310202
28. Kamrin RP, Singer M (1955) The influence of the spinal cord in regeneration of the tail of the lizard, *Anolis carolinensis*. J Exp Zool 128:611–627
29. Maderson PF, Licht P (1968) Factors influencing rates of tail regeneration in the lizard *Anolis carolinensis*. Experientia 24(10):1083–1086
30. Simpson SB Jr (1968) Morphology of the regenerated spinal cord in the lizard, *Anolis carolinensis*. J Comp Neurol 134(2):193–210. doi:10.1002/cne.901340207
31. Turner JE, Singer M (1973) Some morphological and ultrastructural changes in the ependyma of the amputation stump during early regeneration of the tail in the lizard, *Anolis carolinensis*. J Morphol 140(3):257–269. doi:10.1002/jmor.1051400302
32. Zika JM (1969) A histological study of the regenerative response in a lizard, *Anolis carolinensis*. J Exp Zool 172(1):1–8. doi:10.1002/jez.1401720102
33. Bellairs AD, Bryant SV (1985) Autotomy and Regeneration in Reptiles. In: Gans C, Billet F (eds) Biology of the reptilia. Wiley and Sons, New York, pp 301–409
34. Han M, Yang X, Taylor G, Burdsal CA, Anderson RA, Muneoka K (2005) Limb regeneration in higher vertebrates: developing a roadmap. Anat Rec B New Anat 287(1):14–24. doi:10.1002/ar.b.20082
35. Borgens RB (1982) Mice regrow the tips of their foretoes. Science 217(4561):747–750
36. Han M, Yang X, Lee J, Allan CH, Muneoka K (2008) Development and regeneration of the neonatal digit tip in mice. Dev Biol 315(1):125–135. doi:10.1016/j.ydbio.2007.12.025
37. Illingworth CM (1974) Trapped fingers and amputated finger tips in children. J Pediatr Surg 9(6):853–858
38. Singer M, Weckessar EC, Geraudie J, Maier CE, Singer J (1987) Open finger tip healing and replacement after distal amputation in rhesis monkey with comparison to limb regeneration in lower vertebrates. Anat Embryol (Berl) 177(1):29–36
39. Godwin J (2014) The promise of perfect adult tissue repair and regeneration in mammals: Learning from regenerative amphibians and fish. Bioessays 36(9):861–871. doi:10.1002/bies.201300144
40. Lozito TP, Tuan RS (2015) Lizard tail regeneration: regulation of two distinct cartilage regions by Indian hedgehog. Dev Biol 399(2):249–262. doi:10.1016/j.ydbio.2014.12.036

41. Nambiar VV, Bhatt IY, Deshmuhk PA, Jape RR, Jivani PN, Kavale HR, Prakushkar SS, Ramchandran AV (2008) Assessment of extracellular matrix remodeling during tail regeneration in the lizard Hemidactylus flaviviridis. J Endocrinol Reprod 12(2):67–72
42. Poyntz SV, Bellairs A'A (1965) Natural limb regeneration in *Lacerta vivipara*. Br J Herpetol 3:204–205
43. Marcucci E (1930) II potere rigenerativ degli arti nei Rettili. Richerche sperimentale sopra alcune specie di Sauri. Archo Zool Ital 14:27–252
44. Campbell LJ, Crews CM (2008) Woul epidermis formation and function in urodele amphibian limb regeneration. Cell Mol Life Sci 65:73–79. doi:10.1007/s00018-007-7433-z
45. Murawala P, Tanaka EM, Currie JD (2012) Regeneration: the ultimate would healing. Semin Cell Dev Biol 23(9):954–962. doi:10.1016/j.semcdb.2012.09.013
46. Christensen RN, Tassava RA (2000) Apical epithelial cap morphology and fibronectin gene expression in regenerating axolotl limbs. Dev Dyn 217(2):216–224. doi:10.1002/(SICI)1097-0177(200002)217:2<216::AID-DVDY8>3.0.CO;2-8
47. Christensen RN, Weinstein M, Tassava RA (2002) Expression of fibroblast growth factors 4, 8, and 10 in limbs, flanks, and blastemas of *Ambystoma*. Dev Dyn 223(2):193–203. doi:10.1002/dvdy.10049
48. Han MJ, An JY, Kim WS (2001) Expression patterns of Fgf-8 during development and limb regeneration of the axolotl. Dev Dyn 220(1):40–48. doi:10.1002/1097-0177(2000)9999:9999<::AID-DVDY1085>3.0.CO;2-8
49. Stocum DL, Cameron JA (2011) Looking proximally and distally: 100 years of limb regeneration and beyond. Dev Dyn 240(5):943–968. doi:10.1002/dvdy.22553
50. Yokoyama H (2008) Initiation of limb regeneration: the critical steps for regenerative capacity. Dev Growth Differ 50(1):13–22. doi:10.1111/j.1440-169X.2007.00973.x
51. Godwin J, Kuraitis D, Rosenthal N (2014) Extracellular matrix considerations for scar-free repair and regeneration: insights from regenerative diversity among vertebrates. Int J Biochem Cell Biol 56:47–55. doi:10.1016/j.biocel.2014.10.011
52. Vinarsky V, Atkinson DL, Stevenson TJ, Keating MT, Odelberg SJ (2005) Normal newt limb regeneration requires matrix metalloproteinase function. Dev Biol 279(1):86–98. doi:10.1016/j.ydbio.2004.12.003
53. Fahmy GH, Sicard RE (2002) A role for effectors of cellular immunity in epimorphic regeneration of amphibian limbs. In Vivo 16(3):179–184
54. Godwin JW, Brockes JP (2006) Regeneration, tissue injury and the immune response. J Anat 209(4):423–432. doi:10.1111/j.1469-7580.2006.00626.x
55. Godwin JW, Pinto AR, Rosenthal NA (2013) Macrophages are required for adult salamander limb regeneration. Proc Natl Acad Sci U S A 110(23):9415–9420. doi:10.1073/pnas.1300290110
56. Godwin JW, Rosenthal N (2014) Scar-free wound healing and regeneration in amphibians: Immunological influences on regenerative success. Differentiation 87(1–2):66–75. doi:10.1016/j.diff.2014.02.002
57. Harty M, Neff AW, King MW, Mescher AL (2003) Regeneration or scarring: an immunologic perspective. Dev Dyn 226(2):268–279. doi:10.1002/dvdy.10239
58. Mescher AL, Neff AW (2006) Limb regeneration in amphibians: immunological considerations. ScientificWorldJournal 6(Suppl 1):1–11. doi:10.1100/tsw.2006.323
59. Alibardi L (2010) Ultrastructural features of the process of wound healing after tail and limb amputation in lizard. Acta Zool 91(3):306–318. doi:10.1111/j.1463-6395.2009.00409.x
60. Alibardi L, Celeghin A, Valle LD (2012) Wounding in lizards results in the release of beta-defensins at the wound site and formation of an antimicrobial barrier. Dev Comp Immunol 36(3):557–565. doi:10.1016/j.dci.2011.09.012
61. Barron L, Wynn TA (2011) Macrophage activation governs schistosomiasis-induced inflammation and fibrosis. Eur J Immunol 41(9):2509–2514. doi:10.1002/eji.201141869
62. Lucas T, Waisman A, Ranjan R, Roes J, Krieg T, Müller W, Roers A, Eming SA (2010) Differential roles of macrophages in diverse phases of skin repair. J Immunol 184(7):3964–3977. doi:10.4049/jimmunol.0903356

63. Martini R, Fischer S, López-Vales R, David S (2008) Interactions between Schwann cells and macrophages in injury and inherited demyelinating disease. Glia 56(14):1566–1577. doi:10.1002/glia.20766
64. Nucera S, Biziato D, De Palma M (2011) The interplay between macrophages and angiogenesis in development, tissue injury and regeneration. Int J Dev Biol 55(4–5):495–503. doi:10.1387/ijdb.103227sn
65. Delavary BM, van der Veer WM, van Egmond M (2011) Macrophages in skin injury and repair. Immunobiol 216(7):753–762. doi:10.1016/j.imbio.2011.01.001
66. Stefater JA 3rd, Ren S, Lang RA, Duffield JS (2011) Metchnikoff's policemen: macrophages in development, homeostasis and regeneration. Trends Mol Med 17(12):743–752. doi:10.1016/j.molmed.2011.07.009
67. Butler EG, O'Brien JP (1942) Effects of localized X-radiation on regeneration of the urodele limb. Anat Rec (Hoboken) 84(4):407–413. doi:10.1002/ar.1090840408
68. Echeverri K, Tanaka EM (2002) Ectoderm to mesoderm lineage switching during axolotl tail regeneration. Science 298(5600):1993–1996. doi:10.1126/science.1077804
69. Hay ED, Fischman DA (1961) Origin of the blastema in regenerating limbs of the newt *Triturus viridescens*. An autoradiographic study using tritiated thymidine to follow cell proliferation and migration. Dev Biol 3:26–59
70. Iten LE, Bryant SV (1973) Forelimb regeneration from different levels of amputation in the newt, *Notophthalmus viridescens*: length, rate, and stages. Wilhelm Roux Arch Entwickl Mech Org 173(4):263–282. doi:10.1007/BF00575834
71. Mescher AL (1996) The cellular basis of limb regeneration in urodeles. Int J Dev Biol 40(4):785–795
72. Namenwirth M (1974) The inheritance of cell differentiation during limb regeneration in the axolotl. Dev Biol 41(1):42–56
73. Peadon AM, Singer M (1966) The blood vessels of the regenerating limb of the adult newt, *Triturus*. J Morphol 118(1):79–89. doi:10.1002/jmor.1051180106
74. Singer M (1974) Trophic functions of the neuron. VI. Other trophic systems. Neurotrophic control of limb regeneration in the newt. Ann N Y Acad Sci 228:308–322
75. Smith AR, Wolpert L (1975) Nerves and angiogenesis in amphibian limb regeneration. Nature 257(5523):224–225
76. Thornton CS (1938) The histogenesis of the regenerating fore limb of larval *Ambystoma* after exarticulation of the humerus. J Morphol 62(2):219–241. doi:10.1002/jmor.1050620204
77. Wallace BM, Wallace H (1973) Participation of grafted nerves in amphibian limb regeneration. J Embryol Exp Morphol 29(3):559–570
78. Kragl M, Knapp D, Nacu E, Khattak S, Maden M, Epperlein HH, Tanaka EM (2009) Cells keep a memory of their tissue origin during axolotl limb regeneration. Nature 460(7251):60–65. doi:10.1038/nature08152
79. Tanaka EM (2016) The molecular and cellular choreography of appendage regeneration. Cell 165:1598–1608. doi:10.1016/j.cell.2016.05.038
80. Sandoval-Guzmán T, Wang H, Khattak S, Schuez M, Roensch K, Nacu E, Tazaki A, Joven A, Tanaka EM, Simon A (2014) Fundamental differences in dedifferentiation and stem cell recruitment during skeletal muscle regeneration in two salamander species. Cell Stem Cell 14(2):174–187. doi:10.1016/j.stem.2013.11.007
81. Cox PG (1969) Some aspects of tail regeneration in the lizard, *Anolis carolinensis*. I. A description based on histology and autoradiography. J Exp Zool 171(2):127–149. doi:10.1002/jez.1401710202
82. Hughes A, New D (1959) Tail regeneration in the geckonid lizard, *Sphaerodactylus*. J Embryol Exp Morphol 7:281–302
83. Kahn EB, Simpson SB Jr (1974) Satellite cells in mature, uninjured skeletal muscle of the lizard tail. Dev Biol 37(1):219–223
84. Simpson SB Jr (1970) Studies on regeneration of the lizard's tail. Am Zool 10:157–165

85. Firulli BA, Redick BA, Conway SJ, Firulli AB (2007) Mutations within helix I of Twist1 result in distinct limb defects and variation of DNA binding affinities. J Biol Chem 282(37):27536–27546
86. Krawchuk D, Weiner SJ, Chen YT, Lu BC, Costantini F, Behringer RR, Laufer E (2010) Twist1 activity thresholds define multiple functions in limb development. Dev Biol 347(1):133–146. doi:10.1016/j.ydbio.2010.08.015
87. Loebel DA, Hor AC, Bildsoe H, Jones V, Chen YT, Behringer RR, Tam PP (2012) Regionalized twist1 activity in the forelimb bud drives the morphogenesis of the proximal and preaxial skeleton. Dev Biol 362(2):132–140. doi:10.1016/j.ydbio.2011.11.020
88. O'Rourke MP, Tam PP (2002) Twist functions in mouse development. Int J Dev Biol 46(4):401–413
89. Tavares AT, Izpisúja-Belmonte JC, Rodriguez-León J (2001) Developmental expression of chick twist and its regulation during limb patterning. Int J Dev Biol 45(5–6):707–713
90. Zuniga A, Quillet R, Perrin-Schmitt F, Zeller R (2002) Mouse twist is required for fibroblast growth factor-mediated epithelial-mesenchymal signalling and cell survival during limb morphogenesis. Mech Dev 114(1–2):51–59
91. Kragl M, Toensch K, Nusslein I, Tazaki A, Tanguchi Y, Tarui H, Hayashi T, Agata K, Tanaka EM (2013) Muscle and connective tissue progenitor populations show sidtinct Twist1 and Twist3 expression progiles during axolotl limb regeneration. Dev Biol 373(1):196–204. doi:10.1016/j.ydbio.2012.10.019
92. Simpson SB Jr (1964) Analysis of tail regeneration in the lizard Lygosoma laterale. i. Initiation of regeneration and cartilage differentiation: the role of ependyma. J Morphol 114:425–435. doi:10.1002/jmor.1051140305
93. Whimster IW (1978) Nerve supply as a stimulator of the growth of tissues including skin. Clin Exp Dermatol 3(4):389–410
94. Wang Y, Wang R, Jiang S, Zhou W, Liu Y, Wang Y, Gu Q, Gu Y, Dong Y, Liu M, Gu X, Ding F, Gu X (2011) Gecko CD59 is implicated in proximodistal identity during tail regeneration. PLoS One 6(3), e17878. doi:10.1371/journal.pone.0017878.g008
95. Tollis M, Hutchins ED, Kusumi K (2014) Reptile genomes open the frontier for comparative analysis of amniote development and regeneration. Int J Dev Biol 58:863–871. doi: 10.1387/ijdb.140316kk
96. Kim J, Kim J, Kim DW, Ha Y, Ihm MH, Kim H, Song K, Lee I (2010) Wnt5a induces endothelial inflammation via -catenin-independent signaling. J Immunol 185:1274–1282. doi:10.4049/jimmunol.1000181
97. Knapp D, Schulz H, Rascon CA, Volkmer M, Scholz J, Nacu E, Le M, Novozhilov S, Tazaki A, Protze S, Jacob T, Hubner N, Habermann B, Tanaka EM (2013) Comparative transcriptional profiling of the axolotl limb identifies a tripartite regeneration-specific gene program. PLoS One 8(5), e61352. doi:10.1371/journal.pone.0061352.s011
98. Sugiura T, Tazaki A, Ueno N (2009) *Xenopus* Wnt-5a induces an ectopic larval tail at injured site, suggesting a crucial role for noncanonical Wnt signal in tail regeneration. Mech Dev 126(1–2):56–67. doi:10.1016/j.mod.2008.10.002
99. Ghosh S, Roy S, Séguin C, Bryant SV, Gardiner DM (2008) Analysis of the expression and function of Wnt-5a and Wnt-5b in developing and regenerating axolotl (*Ambystoma mexicanum*) limbs. Dev Growth Differ 50(4):289–297. doi:10.1111/j.1440-169X.2008.01000.x
100. Stoick-Cooper CL, Weidinger G, Riehle KJ, Hubbert C, Major MB, Fausto N, Moon RT (2006) Distinct Wnt signaling pathways have opposing roles in appendage regeneration. Development 134(3):479–489. doi:10.1242/dev.001123
101. Wheeler BM, Heimberg AM, Moy VN, Sperling EA, Holstein TW, Heber S, Peterson KJ (2009) The deep evolution of metazoan microRNAs. Evol Dev 11:50–68
102. Stefani G, Slack FJ (2008) Small non-coding RNAs in animal development. Nat Rev Mol Cell Biol 9:219–230
103. Pourrajab F, Babaei Zarch M, BaghiYazdi M, Hekmatimoghaddam S, Zare-Khormizi MR (2014) MicroRNA-based system in stem cell reprogramming; differentiation/dedifferentiation. Int J Biochem Cell Biol 55:318–328

104. Undi RB, Kandi R, Gutti RK (2013) MicroRNAs as haematopoiesis regulators. Adv Hematol 2013:695754
105. Kroesen BJ, Teteloshvili N, Czepiel KS (2015) Immuno-miRs: critical regulators of T-cell development, function and ageing. Immunology 144:1–10. doi:10.1111/imm.12367
106. Williams AH, Liu N, van Rooij E, Olson EN (2009) MicroRNA control of muscle development and disease. Curr Opin Cell Biol 21:461–469
107. Holman EC, Campbell LJ, Hines J, Crews CM (2012) Microarray analysis of microrna expression during axolotl limb regeneration. PLoS One 7, e41804
108. Sehm T, Sachse C, Frenzel C, Echeverri K (2009) miR-196 is an essential early-stage regulator of tail regeneration, upstream of key spinal cord patterning events. Dev Biol 334:468–480
109. Nakamura K, Maki N, Trinh A, Trask HW, Gui J, Tomlinson CR, Tsonis PA (2010) miRNAs in newt lens regeneration: specific control of proliferation and evidence for mirna networking. PLoS One 5, e12058
110. Tsonis PA, Call MK, Grogg MW, Sartor MA, Taylor RR, Forge A, Fyffe R, Goldenberg R, Cowper-Sal-lari R, Tomlinson CR (2007) MicroRNAs and regeneration: let-7 members as potential regulators of dedifferentiation in lens and inner ear hair cell regeneration of the adult newt. Biochem Biophys Res Commun 362:940–945
111. Thatcher EJ, Paydar I, Anderson KK, Patton JG (2008) Regulation of zebrafish fin regeneration by microRNAs. Proc Natl Acad Sci U S A 105:18384–18389
112. Yin VP, Lepilina A, Smith A, Poss KD (2012) Regulation of zebrafish heart regeneration by miR-133. Dev Biol 365:319–327. doi:10.1016/j.ydbio.2012.02.018
113. Yu YM, Gibbs KM, Davila J, Campbell N, Sung S, Todorava TI, Otsuka S, Sabaawy HE, Hart RP, Schachner M (2011) MicroRNA miR-133b is essential for functional recovery after spinal cord injury in adult zebrafish. Eur J Neurosci 33:1587–1597. doi:10.1111/j.1460-9568-2011-x
114. Hutchins ED, Eckalbar WL, Wolter JM, Mangone M, Kusumi K (2016) Differential expression of conserved and novel microRNAs during tail regeneration in the lizard Anolis carolinensis. BMC Genomics 17(1):339–350. doi:10.1186/s12864-016-2640-3

Chapter 3
Dependency on Non-myogenic Cells for Regeneration of Skeletal Muscle

Cherie Alissa Lynch, Alexander B. Andre, and Alan Rawls

3.1 Introduction

In the search to uncover the mechanisms of tissue regeneration and how they can be leveraged for therapeutic approaches, skeletal muscle has become an attractive model. Studies in the genetically tractable mouse have provided insight into the myogenic progenitor cells and signaling networks essential for efficient muscle repair in response to acute and chronic damage. More recently, it has become clear that crosstalk between muscle, the innate immune response and interstitial fibroblastic cells is essential for muscle regeneration. An imbalance in signaling, as observed with chronic inflammation of Duchenne's muscular dystrophy patients, can lead to a progressive increase in fibrosis, fat deposition and muscle necrosis. In contrast, de novo muscle regeneration in response to amputation or severe trauma is largely limited to amphibians, reptiles, and fish among the vertebrates. The additional layers of regulation are necessary to recruit progenitor cells to the site of the amputation as well as impose the positional identity required to accurately regenerate individual muscle groups. Similarly, myeloid and fibroblastic cells have also been shown to participate in these processes. In this chapter, we will review the recent advances in our understanding of the role of non-myogenic cells in muscle regeneration.

C.A. Lynch • A.B. Andre • A. Rawls (✉)
School of Life Sciences, Arizona State University, Life Sciences C Bldg, Tempe, AZ 85287-4501, USA
e-mail: cjlynch1@asu.edu; abandre@asu.edu; Alan.Rawls@asu.edu

J. Wilson-Rawls, K. Kusumi (eds.), *Innovations in Molecular Mechanisms and Tissue Engineering*, Stem Cell Biology and Regenerative Medicine,
DOI 10.1007/978-3-319-44996-8_3

3.2 Satellite Cells of the Myogenic Lineage

Skeletal muscle regeneration is dependent on satellite cells that are functionally defined by their ability to both self-renew and differentiate into myoblasts that are able to fuse to form myofibers. These cells are maintained in a quiescent (G_0 phase) state until environmental cues associated with muscle injury stimulate re-entry into the cell cycle. During effective muscle repair, activated satellite cells migrate to the site of injury, proliferate, and differentiate to generate new muscle fibers.

Satellite cells are characterized by their location beneath the basal lamina of muscle fibers and constitutively express the transcription factors *Pax7* and *Myf5* [1, 2]. Ablation of *Pax7* results in decreased satellite cell proliferation and self-renewal, significantly impacting muscle growth and repair [2]. Quiescent satellite cells (QSCs) have been found to express 500 genes not present in activated satellite cells that participate in cell–cell adhesion, negative regulation of the cell cycle, transcriptional control, and lipid and extracellular matrix transporter activity [3]. Gene loci in QSCs that are only expressed at very low levels until induction via the onset of satellite cell activation are marked by histone H3 Lys4, a marker of active chromatin, indicating that these regions are open, awaiting the signals necessary to prompt activation and begin repair, and not in a dormant state [4, 5]. The ability of QSCs to immediately respond to injury stimuli allows for effective muscle repair.

Upon muscle injury, the myofiber sarcolemma and basal lamina are dismantled, resulting in a disconnection between satellite cells and the collagen-laminin network on which they are anchored. This disruption of the myofiber allows for the release and entry of factors critical for satellite cell activation. One of the first factors implicated in activation, hepatocyte growth factor (HGF), is released from the basal lamina, it then proceeds to bind to the Met receptor on the surface of satellite cells, causing their activation and aiding in their migration to the injury site [6]. Dying fibers within the niche generate nitric oxide (NO), further stimulating HGF release from the basal lamina. Also implicated in the activation and proliferation of satellite cells is the Notch signaling pathway; blockage of Notch leads to inhibition of satellite cell proliferation, whereas up-regulation of Notch leads to the promotion of muscle regeneration [7, 8]. In the muscle niche itself, several factors are secreted that aid in multiple aspects of muscle repair. Fibroblast growth factor (FGF) secretion into the ECM activates the MAPK cascade, resulting in the activation and regulation of satellite cell quiescence [9]. Phosphorylated p38 and MyoD are among the earliest markers of activation, with p38α/β MAPK inducing MyoD protein expression. In support of satellite cell proliferation, Notch3 mRNA and protein levels decline upon activation [10]. Additionally, production of the MYF5 protein begins due to a decrease in miR-31 levels, giving activated satellite cells a *Pax7*$^+$, *Myf5*$^+$ phenotype.

Recently, an additional phase of satellite cell quiescence, termed the G_{alert} phase, has been identified in response to injury. Experiments performed by Rodgers et al. [11], demonstrated that satellite cells residing in muscle in the leg contralateral to the limb with the induced injury were distinct from both quiescent and activated satellite cells. In culture, QSCs in the G_{alert} phase were found to enter the cell cycle earlier than non-injury-induced QSCs. Additionally G_{alert} phase QSCs demonstrated an increase in cell size as compared to QSCs, and a high transcriptional correlation between G_{alert} phase QSCs and activated satellite cells was identified. Both mTORC1 activity and HGF signaling were required for QSCs to switch from G_0 to the G_{alert} phase in response to injury. These findings suggest that G_{alert} phase QSCs retain properties of both QSCs and activated satellite cells in a phase that is "primed" for injury response. In fact, QSCs of the G_{alert} phase demonstrated heightened differentiation in culture and enhanced regeneration following an induced injury in vivo [11].

3.2.1 Proliferation of Satellite Cell and Myoblasts

Satellite cell activation is followed by the rapid expansion of *Pax7*$^+$, *Myf5*$^+$ cells that will form the myoblast population, eventually participating in muscle repair, and self-renewal of a smaller population of *Pax7*$^+$, *Myf5*$^-$ satellite cells that will become quiescent in anticipation of later injury events (Fig. 3.1). The majority of *Pax7*$^+$, *Myf5*$^+$ satellite cells undergo symmetric division, producing two *Pax7*$^+$, *Myf5*$^+$ progenitor cells. WNT7a, acting through its receptors FZD7 and VANGL2, induces symmetric cell division through the planar cell polarity pathway [12]. In addition to HGF, insulin-like growth factor-1 (IGF-1), epidermal growth factor (EGF), transforming growth factors α/β (TGFα and TGFβ), and platelet-derived growth factor (PDGF) also contribute to the proliferation and differentiation of myoblasts [13]. Due to damage of the sarcolemma and basal lamina, myofibers receive an inflow of calcium from the (ECM) matrix, which aids in proteolysis of the myofiber [14]. *Pax*7$^+$, *Myf*5$^+$ cells, stimulated through activated leukocyte secretion of IGF-1 and delivered through capillaries into the niche, will continue to proliferate through the down-regulation of $P27^{kip1}$ and through inactivation of the transcription factor FOXO1 [15]. Negative mitogenic modulation of satellite cells exists through the transforming growth factor β (TGFβ) superfamily, most notably myostatin, which inhibit differentiation of satellite cells through down-regulation of MyoD expression and inhibits activation through the up-regulation of P21 and decreased levels of CDK2 [16, 17]. Tumor necrosis factor α (TNFα) also negatively mediates differentiation through the utilization of the TGFβ activated kinase (TAK1)/p38/NF-kB pathway, resulting in increased levels of Activin A expression to support proliferation [18].

Approximately 10 % of the satellite cell population maintains a *Pax7*$^+$, *Myf5*$^-$ profile and will undergo asymmetrical division to give rise to one *Pax7*$^+$, *Myf5*$^-$ and one *Pax7*$^+$, *Myf5*$^+$ cell (Fig. 3.1). Several signaling pathways present in the microenvironment of the satellite cell niche are responsible for controlling asymmetric satellite cell polarity and fate. Components of the Notch pathway, including a Notch3 effector protein, Notch ligand Delta1 (Dll1), and Notch agonist Numb have all been found to asymmetrically distribute between daughter cells, with

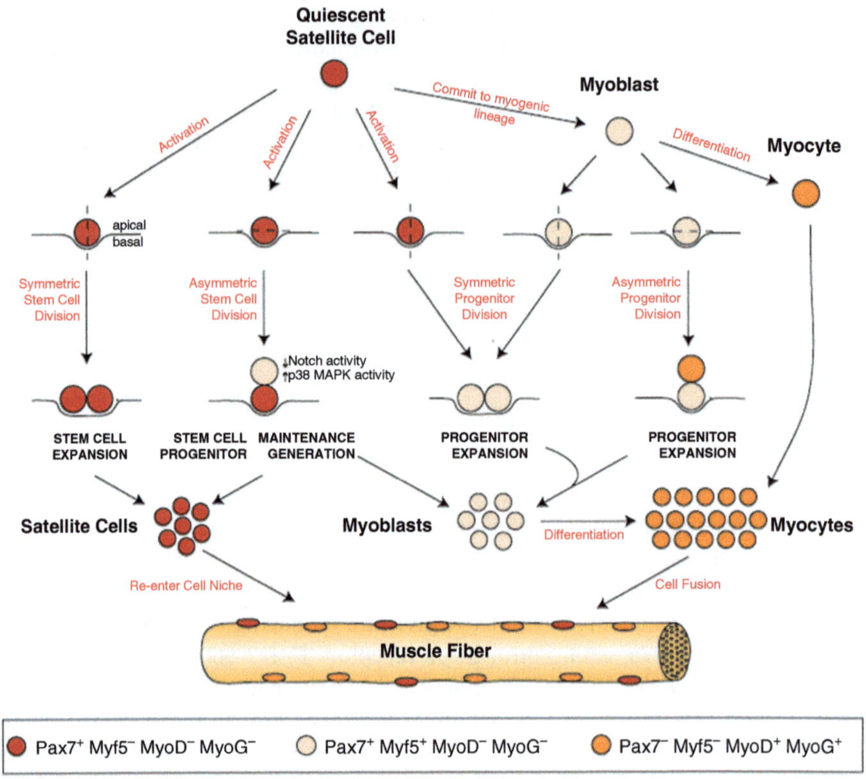

Fig. 3.1 Mechanisms of satellite cell division for muscle maintenance and repair. Following entry to the cell cycle, quiescent satellite cells symmetrically or asymmetrically divide along the apical-basal axis. Symmetric and asymmetric divisions lead to the generation of additional muscle stem cells and progenitor cells. Additionally, satellite cells can directly commit to the myogenic lineage and expand the progenitor cell population or differentiate into myocytes. Resulting muscle stem cells return to the niche to replenish the pool of quiescent satellite cells. Resulting myocytes fuse to form myotubes, leading to the formation of new muscle fibers

DLL1 and NUMB found selectively in the daughter cell committed to becoming a myoblast [8, 19]. Ablation of *Numb* in the muscle lineage profoundly decreased satellite cell proliferation, negatively affecting the ability of muscle to repair following an induced injury [20]. Additionally, factors involved in cell polarity determination, namely parts of the Par complex and Scribbled planar cell polarity protein (Scrib), have been implicated in asymmetric division. Orientation to the myofiber plays an important role in the ability of the satellite cells to asymmetrically divide. This relation to the myofiber, conferred by an apical-basal polarity, is dependent on the interaction of cell membrane receptors basal integrin $\alpha7\beta1$ and apical M-cadherin, resulting in the production of one basal *Pax7*$^{+}$ *Myf5*$^{-}$ cell and one apical *Pax7*$^{+}$ *Myf5*$^{+}$ daughter cell [8]. It has also been proposed that the position of the mitotic spindle in relation to the myofiber axis plays a role in asymmetric division cell fate [21].

3.2.2 Heterogeneity of the Satellite Cell Population

Studies in culture first revealed heterogeneity in the satellite cell population with a "responsive population" that readily proliferates in response to damage and participates in repair, and a "reserve population" that divides at a slow rate and is refractory to differentiation into mature myotubes. This heterogeneity has been reported in muscle tissue at a ratio of 5:1 (responsive: reserve), confirming their relevance to normal muscle biology. The slow dividing cells contribute solely to skeletal muscle when transplanted back into mouse EDL muscle, confirming their commitment to the myogenic lineage. Genome-wide gene expression studies revealed differential expression between the two populations with reserve cells expressing higher levels of inhibitor of differentiation (Id) and other genes that confer "stemness". This predicts that the slow dividing cells that are refractory to repair signals, are essential to muscle homeostasis for long-term maintenance of the satellite cells population.

3.3 Satellite Cell Regulation Through the Stem-Cell Niche

The activation, migration, and proliferation of satellite cells are supported by the inflammatory microenvironment created by components of the niche and immune cells. In addition to ECM, the niche includes fibro-adipogenic (FAP) cells, vasculature, and both residential and infiltrating immune cells that are capable of direct communication with satellite cells. Oxygen free radicals released by neutrophils further break down the sarcolemma, while matrix metalloproteinases released by both damaged myofibers (MMP2) and immune cells (MMP9), aid in the degradation of ECM proteins [22]. ECM digestion through MMPs plays a vital role in satellite cell migration to the site of injury, especially in fibrotic tissue.

FAPs are bipotent fiber-associated cells that also proliferate in response to muscle fiber injury [23]. FAPs double in number in less than 48 h and up-regulate the expression of Interleukin 6 (IL-6) roughly tenfold. IL-6, along with Wnt and IGFs, has been implicated as a pro-differentiation signal that is essential for the differentiation and maturation of myoblasts during muscle repair [23–25]. During myolysis, FAPs have been found to assist in the clearing of cellular debris through phagocytosis of necrotic thymocytes, and when compared to macrophages, FAPs have been found to be fourfold more efficient in debris clearance [26].

Microvasculature and accompanying pericytes help to sustain the cells of the microenvironment, as well as provide the necessary access to circulation for immune cell infiltration in response to damage and delivery of key factors that assist with niche maintenance and satellite cell regulation [4, 17, 27]. PDGF and vascular-endothelial growth factor (VEGF) are released from ruptured blood vessels in response to injury and play an important role in reciprocal communication with satellite cells to promote their proliferation, as well as angiogenesis [28].

Satellite cells are commonly found surrounding the vasculature within a 5 μm radius, with up to 82 % in murine models and 68 % in human residing near capillaries [29]. Pericytes in the muscle serve a jack of all trades role; they help to replace and regenerate the vasculature that can be lost or damaged due to muscle injury, also have been found to replace muscle, and become myogenic in vitro [30]. Pericytes have also been shown to give rise to most of the collagen forming cells during muscle injury, and, in the presence of neurons, have been shown to produce collagens I and III [31].

The ECM contributes to the regulation of satellite cells in the niche. Proteoglycans and glycoproteins play a role in niche homeostasis and in the repair process. Collagen VI ablation in mice leads to a muscle wasting disease not dissimilar to the common dystrophic models [32]. ECM proteins bind to the transmembrane protein dystrophin, forming an anchor that connects the satellite cells to the basal lamina and maintains their anatomical location [33]. ECM proteins can also act as mitogens for satellite cells. Resting, non-damaged satellite cells are located in fibronectin rich regions of the myofiber niche, Syndecan4 (SYN4) and Frizzled7 (FZD7) on the satellite cells act as co-receptors to bind fibronectin [34]. In the presence of WNT7a, this complex will induce symmetrical division. Upon muscle damage, fibronectin is transiently expressed to help maintain the satellite cell pool through the Wnt signaling pathway [34, 35].

The elasticity of the myofiber also plays a role in regulation; normal muscle fibers have a Young's modulus of approximately 12 kPa, while those in aged or dystrophic muscle are much stiffer [36, 37]. This leads to a decrease in quiescent satellite cells because the increased stiffness induces them to enter the cell cycle. Recent work using collagen based scaffolds with elasticity from 2 to 25 kPa as determined by atomic force microscopy (AFM), has shown that on substrates that measure 2 kPa most of the satellite cells maintain their quiescent states and do not enter the cell cycle. Whereas at 25 kPa only about 45 % remain quiescent in vitro [38]. These findings could explain why in aged or dystrophic muscle there is a decreased satellite cell presence, as these two niche environments have an increased stiffness [38, 39].

3.4 Innate Immune Response During Skeletal Muscle Repair

Regeneration of skeletal muscle cannot be accomplished solely by satellite cells. Several types of immune cells, both resident and infiltrating, play an indispensable role in effective tissue regeneration. In healthy homeostatic muscle, immune cells are kept at a minimum, however, disruption of the basal lamina and sarcolemma of myofibers initiates several waves of immune cell infiltration that play discrete roles in the removal of necrotic fibers, activation of satellite cells, and ultimately the efficient differentiation into mature muscle fibers. The majority of the immune cells involved in muscle repair are those of the innate leukocyte lineage—macrophages, neutrophils, dendritic cells, mast cells, eosinophils, basophils, and natural killer cells. Central to the innate immune response is the production and responsiveness to cytokines, chemokines, and growth factors. These signaling molecules mediate crosstalk with satellite cells and FAP cells during the repair process.

Immediately upon myofiber damage, resident mast cells within the muscle degranulate, releasing TNFα, while resident macrophages release C-C motif chemokine ligand 2 (CCL2) and C-X-C motif chemokine ligand 3 (CXCL3), recruiting transient polymorphonuclear neutrophils from the circulation to the site of injury [40]. Satellite cells also contribute to chemoattraction to the site of damage through the release of the pro-inflammatory cytokines IL-1, IL-6, and TNFα [41]. Neutrophils rapidly invade the injured tissue in significant numbers and persist in the tissue for approximately 24 h, where they promote sarcolemma damage through the release of oxygen-free radicals [42]. Through the secretion of IL-1 and IL-8, neutrophils promote the recruitment of circulating $CX3CR1^{low}$, $Ly6C^{+}$, $CCR2^{+}$ phenotype monocytes to the site of injury [43] and binding of CCL2 and CCL7, by the C-C motif chemokine receptor, CCR2 [44]. Disruption of either receptor or ligands leads to severe deficits in monocyte recruitment and efficient muscle repair [45–47]. The infiltrating monocytes differentiate into macrophage subtypes, both pro- and anti-inflammatory, in a process that is highly dependent on the tissue microenvironment.

At approximately 24-h post muscle injury, monocytes/macrophages begin to express high levels of IL-6, supporting macrophage infiltration and myoblast proliferation through the STAT3 pathway. Effective muscle repair requires sufficient generation of myoblasts for regeneration of the damaged tissue. Knockout of IL-6, or knockdown of STAT3, resulted in decreased MyoD, Myogenin, and macrophage infiltration, ultimately resulting in diminished muscle repair [48].

Initially, the pro-inflammatory phenotype is maintained as neutrophils secrete Th1 inflammatory cytokines, interferon-gamma (IFNγ) and TNFα, to induce monocytes to polarize into M1 macrophages ($CX3CR1^{low}$, $Ly6C^{+}$, $CCR2^{+}$). In addition to IFNγ and TNFα, pathogens and granulocyte macrophage colony-stimulating factor (GM-CSF) are capable of stimulating M1 macrophage polarization [49] (Fig. 3.2). M1 macrophages phagocytose cellular debris and secrete factors, such as IL-1b and IL-12, to recruit additional inflammatory cells for debris clearance and pathogen removal. Nitric oxide (NO), produced by M1 cells acts to lyse cells for removal, however, if dysregulated, it can lead to increased tissue damage [50]. During the pro-inflammatory phase, which occurs approximately 24–96 h post injury, the NF-kB pathway in both macrophages and myoblasts is activated in response to TNFα. In macrophages, this enhances the inflammatory response by stimulating the release additional pro-inflammatory cytokines. In muscle, CyclinD1 expression is induced, while MyoD expression is suppressed, in response to activation of the NF-kB pathway, supporting myoblast proliferation and preventing differentiation [51, 52].

Phagocytosis by M1 macrophages and exposure to CSF-1 induce macrophage polarization to skew from a pro-inflammatory phenotype towards an anti-inflammatory phenotype, resolving the inflammation and beginning the muscle repair process [53]. Infiltrating monocytes now become $CX3CR1^{hi}$, $Ly6C^{-}$, $CCR2^{-}$

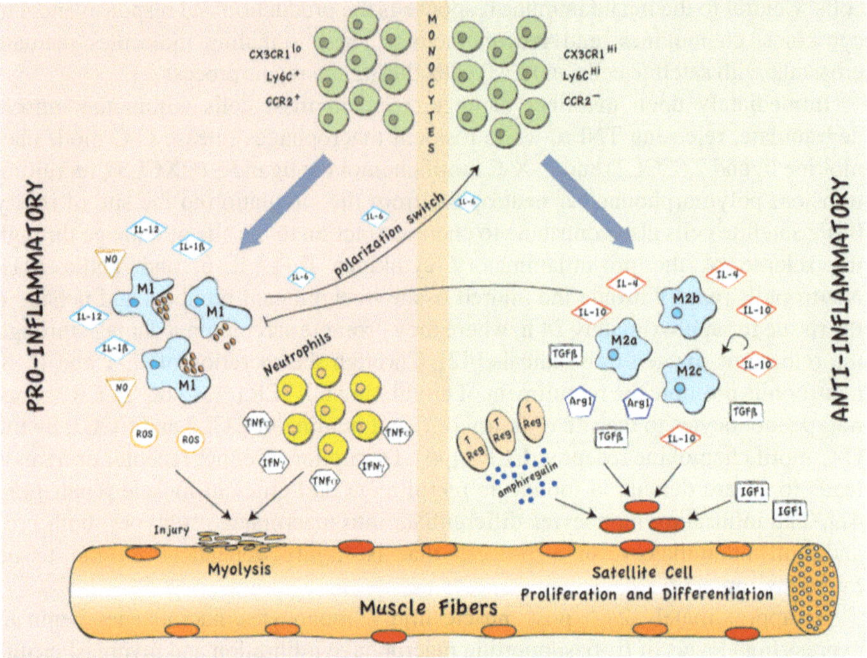

Fig. 3.2 Immune cell contribution and modulation in damaged muscle tissue. In response to myofiber injury, neutrophils from circulation invade the site of damage where they aid in further tissue break down and recruit CX3CR1^{Lo}, Ly6C^{+}, CCR2^{+} monocytes, differentiating into M1 macrophages, for continued debris clearance and pro-inflammatory cytokine secretion. M1 phagocytosis induces macrophage polarization towards an anti-inflammatory phenotype to support muscle repair. CX3CR1^{Hi}, Ly6C^{Lo}, CCR2^{-} monocytes differentiate into M2a, M2b, and M2c macrophages, functioning to suppress inflammation and promote satellite cell proliferation and differentiation. T regulatory cells assist M2 macrophages in resolving inflammation and fostering muscle repair

and differentiate into three subtypes of M2 macrophages. Several molecules have been identified as regulators of the switch from early pro-inflammatory to late anti-inflammatory macrophage phenotypes. cAMP response element-binding protein (CREB), a multifunctional transcription factor, is critical for the up-regulation of genes associated with M2 macrophages (IL-10, IL-13R, Arg-1) and repression of M1 macrophage activation [54]. Mitogen-activated protein kinase (MAPK) phosphatase-1, through inhibition of p38 MAPK activation, functions to control macrophage subtype shifting. MAPK also helps to resolve inflammation to allow for proper muscle repair [55]. Recently, AMP-activated protein kinase (AMPK), widely known as a regulator of metabolic homeostasis, has also been identified as a regulator of macrophage polarization skewing. Mounier et al. [56], demonstrated loss of M2 macrophage functionality and a loss of M2 markers expressed in AMPKα1$^{-/-}$ macrophages. Further, AMPKα1$^{-/-}$ mice showed deficient muscle repair resulting from a failure of M1 macrophage phagocytosis-induced polarization to an M2 phenotype [56].

M2a macrophages arise from the release of IL-4 or IL-13 and signal via IL-4 receptor alpha [57]. Release of these Th2 inflammatory cytokines causes increased expression of CD206 and CD36 by macrophages. In vitro, it has been shown that M2a macrophages, producing arginase, decrease M1 macrophage lysis activity through competition for arginine, the shared enzymatic substrate of arginase and iNOS [58]. M2a macrophages secrete IL-10 and TGF-β, thereby inducing the anti-inflammatory M2c macrophage subtype, which aids in IL-10 and TGF-β release (Fig. 3.2). Secretion of these cytokines suppresses inflammation and promotes satellite cell proliferation, allowing for remodeling of the extracellular matrix, angiogenesis, and muscle fiber development to begin [58]. Glucocorticoids and IFNβ can also stimulate the induction of the M2c subtype [59]. The release of IL-4 by M2b regulatory macrophages, Th2 cells, eosinophils, and basophils further promotes the wound healing phase by decreasing phagocytosis and stimulating macrophage fusion [49]. In addition to IL-4, the release of IGF-1 also contributes to continued satellite cell growth and myofiber fusion [60]. In recent experiments by Tonkin et al. [61], macrophages were identified as a major contributing source of IGF-1 at the site of muscle damage. Indeed, when muscle injury is induced in mice devoid of IGF-1 in myeloid cells, a loss of regenerative capacity is demonstrated. During the late stages of healthy muscle repair, $Ly6C^+$ monocytes/macrophages and $CD206^+$ macrophages were found to express high levels of IGF-1. However, when IGF-1 is knocked out from myeloid cells, the population of $Ly6C^+$ monocytes/macrophages is heightened while the population of $CD206^+$ macrophages is diminished [61].

Aiding in the establishment of the anti-inflammatory environment at the site of muscle damage, a population of $CD4^+$ regulatory T cells (T_{reg}) arises concurrently with M2 macrophages, though to a much lesser extent (Fig. 3.2). FoxP3, a forkhead transcription factor, regulates T_{reg} cell lineage specification, however, it remains unclear whether the population of T_{reg} cells at the site of muscle injury derives from resident T_{reg} cells in the muscle or is recruited in response to damage. T_{reg} cells have been shown to influence myeloid and T cell infiltration, as well as satellite cell colony-forming capacity. Additionally, T_{reg} cells were found express IL-10 and amphiregulin, which accumulate during the final stages of muscle repair and play important roles in negative regulation of inflammation and satellite cell activation and proliferation, respectively [62]. Due to the capability of T_{reg} cells to modulate the inflammatory response and satellite cell activity, research in using T_{reg} cells to improve muscle repair is of current interest. Villalta et al. demonstrated increased levels of T_{reg} cells in both human Duchenne's muscular dystrophy (DMD) and in the corresponding *mdx* mouse. When T_{reg} cells are depleted from dystrophic muscle, a heightened Th1-cell-mediated response occurs causing increased myofiber damage [63].

In recent years, the multi-faceted role of macrophages in wound repair has begun to lend itself to potential use in therapy for muscle injury. M1-polarized macrophages delivered to the site of muscle damage resulted in enhanced recovery of functionality with reduced myofiber damage and collagen accumulation [64]. When M2a or M2c macrophages are injected, an increase in tube-like structures is observed, indicating improved angiogenesis [65]. To further aid in the repair of muscle injury, especially in cases of volumetric muscle loss, tissue scaffolds with inert or biodegradable properties have been the predominating focus. Contrary to avoiding an immune response, recent work has sought to take advantage of immune cells in the delivery of tissue scaffolds—now termed "smart scaffolds". Macrophages and other inflammatory cells, such as cytokines capable of modulating macrophage polarization, can be loaded into tissue scaffolds prior to transplantation, allowing for a therapeutic approach that is personalized and works in conjunction with the patient's own immune response to enhance the repair process. Through an injectable multidomain peptide scaffold engineered by Kumar et al. the potential to recruit specific inflammatory cells and deliver cytokines to the site of injection was shown. MCP-1 and IL-4 loaded hydrogel scaffolds were capable of boosting macrophage recruitment and stimulating polarization towards a pro-healing M2 phenotype in a time-controlled manner, without inducing a local inflammatory response [66].

3.5 De Novo Regeneration of Skeletal Muscle

As described above, mammalian models have been powerful tools in parsing the signaling pathways regulating the regeneration of skeletal muscle in response to acutely damaged muscle. However, de novo muscle regeneration in response to amputation is largely limited to amphibians, reptiles and fish among the vertebrates. This process can be distinguished by the additional layers of regulation necessary to recruit progenitor cells to the site of the amputation and a complex set of temporal and spatial signals necessary to impose the positional identity required to accurately recapitulate individual muscle groups and coordinate the regeneration of distinct cell lineages that give rise to the skeletal elements, connective tissue, nerves, vasculature, and skin [67]. As with tissue repair, the study of skeletal muscle regeneration has been central to our understanding of complex tissue regeneration. Non-myogenic cell types have been implicated in this process. In this section, we will compare the regulation of muscle repair to regeneration through the lens of the microenvironment created by the immune cells and myofibroblasts.

3.5.1 Amphibians as a Model for the Study of Skeletal Muscle Regeneration

Members of the Anura (frogs and toads) and Caudata (salamanders and newts) orders are the most commonly studied amphibians for muscle regeneration. Anurans possess distinct developmental windows preceding metamorphosis where complete regeneration of organs can occur, while the urodeles (Caudata) are able to regenerate a wide variety of organs throughout adulthood. Perhaps the best studied regenerative tissue system has been limb and tail amputations that follow a conserved set of temporal events that include (1) a modified wound healing process, (2) progenitor cell recruitment and (3) activation and tissue rebuilding (reviewed in [67–69]). Conserved regulatory pathways shared between amphibian models has provided insight into how regeneration has been maintained in these animals and largely lost in mammals.

3.5.2 Wound Healing and ECM Remodeling During Regeneration

Wound healing associated with regeneration shares many common features with scar-free wound healing associated with skin repair. Within hours of amputation, epithelial cells and dermal fibroblasts migrate to the site of injury and cover the fibrin blood clot. The regenerative epithelial cells thicken to form an apical ectodermal cap (AEC) reminiscent of the apical ectodermal ridge (AER) that appears during limb development. The AEC promotes the remodeling of the basement membrane ECM through recruitment of leukocytes and the release growth factors that are capable of inducing the subjacent mesenchymal cells to form a blastema of undifferentiated proliferating progenitor cells with the ability to rise to the distinct cell types of the limb [70–72]. In the case of skeletal muscle, progenitor cells can be derived from myoblasts (*Pax7*$^-$, *MyoG*$^+$) that dedifferentiate muscle fibers and aid in the recruitment of satellite cells (*Pax7*$^+$, *MyoG*$^-$) [73, 74].

The ECM at the site of the wound is recognized as an important regulator of wound healing and the progression towards regeneration. ECM is a complex network of proteins composed primarily of collagens, laminins and fibronectins that interact to create scaffolding as well as serve as adhesion sites for cells through integrin binding. Small leucine-rich proteoglycans within the ECM bind growth factors and cytokines that create microenvironment niches for cell signaling [75]. Within hours of amputation, migrating epithelial cells express matrix metalloproteinases (MMP) that promote ECM breakdown through the digestion of collagen. This facilitates cell invasion, debris clearance and release of the growth factors and cytokines that promote cell migration [72, 76]. A second wave of MMP expression after 3 days is believed to participate in ECM remodeling and promoting muscle dedifferentiation [77]. Treating newt wounds with MMP inhibitors resulted in shortened stumps with distal scars, indicating the importance of the ECM remodeling during regeneration [78]. Macrophages represent important regulators of ECM breakdown and remodeling at the wound site. Inflammatory cytokines produced by macrophages regulate ECM production from fibroblasts and myofibroblasts and ensure a pro-regenerative microenvironment at the site of the wound instead of an acellular fibrotic scar [79, 80]. Depletion of macrophages in salamanders inhibits limb regeneration and promotes the formation of a distal scar and an overrepresentation of myofibroblasts [81]. This underscores the important relationship between the organism's ability to remodel ECM and the formation of fibrotic scars that prevent regeneration. In support of this, salamanders maintain the expression of other developmentally regulated collagens III and XII, tenascin, and hyaluronic acid later into adulthood than mice and delay the onset of collagen I that gives rise to acellular scars through cross-linking with heparin sulfate proteoglycans [81, 82].

3.5.3 Myogenic Progenitor Cells During Regeneration

In classic experiments initially performed in salamanders, myogenic progenitor cells contributing to the blastema were found to be derived through the dedifferentiation of injured muscle [83–85]. Dedifferentiation is characterized by a loss of differentiated muscle-specific markers, fragmentation of multinucleated myotubes into mononucleated cells and re-entry into the cell cycle [86]. The resultant mononucleated Pax7$^-$ MyoG$^+$ cells are capable of redifferentiation into muscle [87]. Cre-loxP-based genetic fate mapping experiments have demonstrated that cells generated through dedifferentiation remain restricted to the myogenic lineage and are unable to contribute to other tissues of the limb or tail [68, 88].

Several transcription factors and cell cycle regulators have been shown to regulate muscle dedifferentiation [74, 89–91]. Perhaps the best studied are members of the MSX family of the homeodomain-containing transcription factors (MSX1 and MSX2) that have been implicated in maintaining cells in proliferative, progenitor state during limb development across vertebrates. Over expression of either MSX1 or MSX2 is sufficient to drive myotube dedifferentiation in culture and the formation of differentiation-competent myoblasts [90]. More recently, it was found that the LIM homeobox transcription factor, *Lhx2*, which can suppress muscle-specific transcription and differentiation in C2C12 cells, is a direct regulator of Msx1 and Msx2 transcription [92]. Further, ectopic expression of MSX1 or MSX2 can induce dedifferentiation of mammalian myotubes suggesting the elements of the dedifferentiation regulatory network of the amphibians have been retained in mammals [93–95].

Inactivation of the tumor suppressor Retinoblastoma (Rb) through phosphorylation has also been implicated in muscle regeneration in the newt limb, consistent with the requirement for reinitiating the cell cycle during generating progenitor cells [74]. Inactivation of Rb is sufficient to promote DNA synthesis in differentiated mouse muscle in culture, however, the cells will not progress to proliferating myoblasts with the capacity for redifferentiation [96, 97]. Complete recapitulation of the dedifferentiation pathway requires an additional insult to the p53 signaling pathway through inactivation of the Alternate Reading Frame (ARF) of the *Ink4a* locus [91]. Interestingly, the earliest identified *ARF* ancestor is in chickens, with no candidates in databases for non-amniote organisms [98–100]. This raises the possibility that loss of regenerative capacity in mammals is related to acquisition of additional levels of cell cycle regulation. There is evidence that environmental cues participate in the regulation of muscle fiber dedifferentiation. The ECM in the tissue proximal to the site of amputation undergoes a shift from a collagen and laminin-based stiff ECM to a softer transitional ECM rich in hyaluronic acid, tenascin-C and fibronectin. Under cell culture conditions, this ECM differentially directs DNA synthesis, migration, myotube fragmentation and myoblast fusion [101, 102].

In addition to the generation of *Pax7*$^-$, *Myog*$^+$ myoblasts through dedifferentiation, there is evidence that recruitment of *Pax7*$^+$, *Myog*$^-$ satellite cells from muscle proximal to the site of amputation participates in muscle regeneration in salamanders [73]. Further, cultured satellite cells are able to contribute to muscle regeneration upon transplantation [68, 103]. This indicates that the system for recruiting myogenic progenitor cells in mammals can participate in regeneration in amphibians as well. Cre-loxP-based genetic fate mapping approaches have been used to track cells in the blastema that are *Pax7*$^-$, *Myog*$^+$ and *Pax7*$^+$, *Myog*$^-$ [68, 88]. Surprisingly, there was a preference for the recruitment of a premyogenic cell source between urodeles, with the *Notophthalmus viridescens* (newt) depending on dedifferentiation of muscle while the *Ambystoma mexicanum* (axolotl) leverages satellite cells [88]. The newt employs a dedifferentiation strategy for the regeneration of other tissues, including the lens of the eye, while the axolotl has limited regenerative capacity for the lens [104, 105]. This reveals a divergence in strategies for generating progenitor cells for tissue of two urodeles separated by approximately 100 million years. This raises interesting questions about the evolutionary pressures that would maintain two discrete mechanisms. The selection process has been strong enough that mammalian muscle is able to functionally recapitulate dedifferentiation with relatively small changes in gene expression of extracellular matrix.

3.5.4 Role of Pro- and Anti-Inflammatory Immune Response in Regeneration

The duality of the innate immune response with the pro-inflammatory arm directed by Th1 cytokines and the anti-inflammatory arm directed by Th2 cytokines is conserved in urodeles. However, analysis of the cytokines post limb amputation reveals two overlapping spikes in Th1 and Th2 cytokines as well as CCL and CXCL chemokines at days 2 and 7, which predicts that anti-inflammatory M2 macrophages are recruited concurrently to the site of injury with pro-inflammatory M1 macrophages [81]. This is in contrast to mammalian muscle repair, where a distinct early wave of pro-inflammatory M1 macrophages is followed by anti-inflammatory M2 macrophages. The presence of M2 macrophages and Th2 cytokines did not inhibit the phagocytic activity of M1 macrophages in the first 24 h post-amputation in the salamander, suggesting a different functional relationship between the two cell types during regeneration. Interestingly, M1 macrophage activity requires expression of anti-inflammatory cytokines as well as several signalling pathways critical for regeneration, including metalloproteinases MMP9 and MMP3, dedifferentiation regulator *Msx2*, blastemal markers *Prrx1* and *Sp9*, the production of Th2 cytokines, and TGFβ signaling [81]. Thus, despite the temporal overlap, modulation of the pro-inflammatory immune response is essential for promoting regeneration.

Studies in Anurans, where regenerative capacity is limited to a pre-metamorphosis time period provides an opportunity to compare cellular processes associated with repair in permissive and non-permissive stages to examine mechanisms by which the immune system regulates regeneration (reviewed in [106]). *Xenopus*, the most common anuran model, will undergo complete limb or tail regeneration between pre-metamorphosis stages 50–53. After metamorphosis has started (stages 57–60), regeneration is only partially complete as exemplified by a cartilaginous spike replacing an amputated limb. The shift from tadpole to adult is associated with immunological shifts from a relatively simple "ancestral" system to one that is more complex and resembles that of the mammals [107, 108]. Consistent with this, differential gene expression studies between regeneration competent and incompetent stages confirms differences in the immune signaling and resolution of inflammation [109–111]. While pro-inflammatory signals spike early after limb amputation in stage 53 of *Xenopus*, they persist at the regeneration non-competent stage 57 [110]. This would indicate that unresolved inflammation in response to injury contributes to the loss of the regenerative capacity in adult frogs. In support of this, immune cell depletion can extend the period of regeneration competence in *Xenopus* [112].

Studies in anurans and urodeles have provided seemingly conflicting models of the role of the inflammatory response to regeneration, with disruption of inflammatory macrophages inhibiting salamander and newt regeneration while extending the regenerative refractory period in frogs [81, 112]. This can best be reconciled through the lens of comparative strength of the immune system. Salamanders are considered to have a strong innate immune system, but because of the lack of key adaptive immune responses, it is considered relatively weak compared to the frog and mouse [113]. In the case of the frog, the strength of the immune system increases with age, leading to the hypothesis that the regenerative capacity of the organism is inversely proportional to the strength of the immune response to injury. This is likely an over-simplified axiom as phagocytotic macrophages are essential for salamander limb and tail regeneration. There has been considerable effort to understand the immune response to pathogens and this can provide insight into differences in humoral and cytotoxic immune response between amphibians [72, 114, 115]. Understanding how the broader immune system plays a role in tissue regeneration should help resolve this confusion.

References

1. Cornelison DD, Wold BJ (1997) Single-cell analysis of regulatory gene expression in quiescent and activated mouse skeletal muscle satellite cells. Dev Biol 191:270–283. doi:10.1006/dbio.1997.8721
2. Seale P, Sabourin LA, Girgis-Gabardo A et al (2000) Pax7 is required for the specification of myogenic satellite cells. Cell 102:777–786. doi:10.1016/S0092-8674(00)00066-0
3. Shea KL, Xiang W, LaPorta VS et al (2010) Sprouty1 regulates reversible quiescence of a self-renewing adult muscle stem cell pool during regeneration. Cell Stem Cell 6:117–129. doi:10.1016/j.stem.2009.12.015
4. Fukada S, Uezumi A, Ikemoto M et al (2007) Molecular signature of quiescent satellite cells in adult skeletal muscle. Stem Cells 25:2448–2459. doi:10.1634/stemcells.2007-0019
5. Guenther MG, Levine SS, Boyer LA et al (2007) A chromatin landmark and transcription initiation at most promoters in human cells. Cell 130:77–88. doi:10.1016/j.cell.2007.05.042

6. Tatsumi R, Anderson JE, Nevoret CJ et al (1998) HGF/SF is present in normal adult skeletal muscle and is capable of activating satellite cells. Dev Biol 194:114–128. doi:10.1006/dbio.1997.8803
7. Conboy IM, Conboy MJ, Smythe GM, Rando TA (2003) Notch-mediated restoration of regenerative potential to aged muscle. Source Sci New Ser 302:1575–1577. doi:10.1126/science.1087573
8. Kuang S, Kuroda K, Le Grand F, Rudnicki MA (2007) Asymmetric self-renewal and commitment of satellite stem cells in muscle. Cell 129:999–1010. doi:10.1016/j.cell.2007.03.044
9. Jones NC, Tyner KJ, Nibarger L et al (2005) The p38α/β MAPK functions as a molecular switch to activate the quiescent satellite cell. J Cell Biol 169:105–116. doi:10.1083/jcb.200408066
10. Mourikis P, Sambasivan R, Castel D et al (2012) A critical requirement for notch signaling in maintenance of the quiescent skeletal muscle stem cell state. Stem Cells 30:243–252. doi:10.1002/stem.775
11. Rodgers JT, King KY, Brett JO et al (2014) mTORC1 controls the adaptive transition of quiescent stem cells from G0 to G(Alert). Nature 509:393–396. doi:10.1038/nature13255
12. Le Grand F, Jones AE, Seale V et al (2009) Wnt7a activates the planar cell polarity pathway to drive the symmetric expansion of satellite stem cells. Cell Stem Cell 4:535–547. doi:10.1016/j.stem.2009.03.013
13. Wozniak AC, Anderson JE (2007) Nitric oxide-dependence of satellite stem cell activation and quiescence on normal skeletal muscle fibers. Dev Dyn 236:240–250. doi:10.1002/dvdy.21012
14. Oberc MA, Engel WK (1977) Ultrastructural localization of calcium in normal and abnormal skeletal muscle. Lab Invest 36(6):566–577, PMID: 17033
15. Chakravarthy MV, Davis BS, Booth FW (2000) IGF-I restores satellite cell proliferative potential in immobilized old skeletal muscle. J Appl Physiol 89:1365–1379. doi:10.1111/j.1600-0838.2001.110111-3.x
16. Langley B, Thomas M, Bishop A et al (2002) Myostatin inhibits myoblast differentiation by down-regulating MyoD expression. J Biol Chem 277:49831–49840. doi:10.1074/jbc.M204291200
17. McCroskery S, Thomas M, Maxwell L et al (2003) Myostatin negatively regulates satellite cell activation and self-renewal. J Cell Biol 162:1135–1147. doi:10.1083/jcb.200207056
18. Trendelenburg AU, Meyer A, Jacobi C et al (2012) TAK-1/p38/nNFkB signaling inhibits myoblast differentiation by increasing levels of Activin A. Skelet Muscle 2:3. doi:10.1186/2044-5040-2-3
19. Conboy IM, Rando TA (2002) The regulation of notch signaling controls satellite cell activation and cell fate determination in postnatal myogenesis. Dev Cell 10:273. doi:10.1016/j.devcel.2006.01.003
20. George RM, Biressi S, Beres BJ et al (2013) Numb-deficient satellite cells have regeneration and proliferation defects. Proc Natl Acad Sci U S A 110:18549–18554. doi:10.1073/pnas.1311628110
21. Kuang S, Gillespie MA, Rudnicki MA (2008) Niche regulation of muscle satellite cell self-renewal and differentiation. Cell Stem Cell 2:22–31. doi:10.1016/j.stem.2007.12.012
22. Kherif S, Lafuma C, Dehaupas M et al (1999) Expression of matrix metalloproteinases 2 and 9 in regenerating skeletal muscle: a study in experimentally injured and mdx muscles. Dev Biol 205:158–170. doi:10.1006/dbio.1998.9107
23. Joe AW, Yi L, Natarajan A et al (2010) Muscle injury activates resident fibro/adipogenic progenitors that facilitate myogenesis. Nat Cell Biol 12:153–163. doi:10.1038/ncb2015
24. Otto A, Schmidt C, Luke G et al (2008) Canonical Wnt signalling induces satellite-cell proliferation during adult skeletal muscle regeneration. J Cell Sci 121:2939–2950. doi:10.1242/jcs.026534
25. Serrano AL, Baeza-Raja B, Perdiguero E et al (2008) Interleukin-6 is an essential regulator of satellite cell-mediated skeletal muscle hypertrophy. Cell Metab 7:33–44. doi:10.1016/j.cmet.2007.11.011
26. Heredia JE, Mukundan L, Chen FM et al (2013) Type 2 innate signals stimulate fibro/adipogenic progenitors to facilitate muscle regeneration. Cell 153:376–388. doi:10.1016/j.cell.2013.02.053

27. Machida S, Booth FW (2004) Insulin-like growth factor 1 and muscle growth: implication for satellite cell proliferation. Proc Nutr Soc 63:337–340. doi:10.1079/PNS2004354
28. Montarras D, L'Honoré A, Buckingham M (2013) Lying low but ready for action: the quiescent muscle satellite cell. FEBS J 280:4036–4050. doi:10.1111/febs.12372
29. Christov C, Chrétien F, Abou-Khalil R et al (2007) Muscle satellite cells and endothelial cells: close neighbors and privileged partners. Mol Biol Cell 18:1397–1409
30. Dellavalle A, Sampaolesi M, Tonlorenzi R et al (2007) Pericytes of human skeletal muscle are myogenic precursors distinct from satellite cells. Nat Cell Biol 9:255–267. doi:10.1038/ncb1542
31. Dulauroy S, Di Carlo SE, Langa F et al (2012) Lineage tracing and genetic ablation of ADAM12$^+$ perivascular cells identify a major source of profibrotic cells during acute tissue injury. Nat Med 18:1262–1270. doi:10.1038/nm.2848
32. Bönnemann CG (2007) The collagen VI-related myopathies: muscle meets its matrix. Nat Rev Neurol 7:379–390. doi:10.1038/nrneurol.2011.81
33. Michele DE, Campbell KP (2003) Dystrophin-glycoprotein complex: post-translational processing and dystroglycan function. J Biol Chem 278:15457–15460. doi:10.1074/jbc.R200031200
34. Bentzinger CF, Wang YX, Von Maltzahn J et al (2013) Fibronectin regulates Wnt7a signaling and satellite cell expansion. Cell Stem Cell 12:75–87. doi:10.1016/j.stem.2012.09.015
35. Han WM, Jang YC, García AJ (2016) Engineered matrices for skeletal muscle satellite cell engraftment and function. Matrix Biol 53:1–14. doi:10.1016/j.matbio.2016.06.001
36. Engler AJ, Griffin MA, Sen S et al (2004) Myotubes differentiate optimally on substrates with tissue-like stiffness: pathological implications for soft or stiff microenvironments. J Cell Biol 166:877–887. doi:10.1083/jcb.200405004
37. Gao Y, Kostrominova T (2008) Age-related changes in the mechanical properties of the epimysium in skeletal muscles of rats. J Biomech 41:465–469
38. Quarta M, Brett JO, DiMarco R et al (2016) An artificial niche preserves the quiescence of muscle stem cells and enhances their therapeutic efficacy. Nat Biotechnol 34(7):752–759. doi:10.1038/nbt.3576
39. Boontheekul T, Hill EE, Kong H-J, Mooney DJ (2007) Regulating myoblast phenotype through controlled gel stiffness and degradation. Tissue Eng 13:1431–1442. doi:10.1089/ten.2006.0356
40. Brigitte M, Schilte C, Plonquet A et al (2010) Muscle resident macrophages control the immune cell reaction in a mouse model of notexin-induced myoinjury. Arthritis Rheum 62:268–279. doi:10.1002/art.27183
41. Tews DS, Goebel HH (1996) Cytokine expression profile in idiopathic inflammatory myopathies. J Neuropathol Exp Neurol 55:342–347. doi:10.1017/CBO9781107415324.004
42. Tidball JG (1995) Inflammatory cell response to acute muscle injury. Med Sci Sports Exerc 27(7):1022–1032
43. Saini J, McPhee JS, Al-Dabbagh S et al (2016) Regenerative function of immune system: modulation of muscle stem cells. Ageing Res Rev 27:67–76. doi:10.1016/j.arr.2016.03.006
44. Tsou CL, Peters W, Si Y et al (2007) Critical roles for CCR2 and MCP-3 in monocyte mobilization from bone marrow and recruitment to inflammatory sites. J Clin Invest 117:902–909. doi:10.1172/JCI29919
45. Lu H, Huang D, Ransohoff RM, Zhou L (2011) Acute skeletal muscle injury: CCL2 expression by both monocytes and injured muscle is required for repair. FASEB J 25:3344–3355. doi:10.1096/fj.10-178939
46. Warren GL, Hulderman T, Mishra D et al (2005) Chemokine receptor CCR2 involvement in skeletal muscle regeneration. FASEB J 19:413–415. doi:10.1096/fj.04-2421fje
47. Warren GL, O'Farrell L, Summan M et al (2004) Role of CC chemokines in skeletal muscle functional restoration after injury. Am J Physiol Cell Physiol 286:C1031–C1036. doi:10.1152/ajpcell.00467.2003
48. Zhang C, Li Y, Wu Y et al (2013) Interleukin-6/signal transducer and activator of transcription 3 (STAT3) pathway is essential for macrophage infiltration and myoblast proliferation during muscle regeneration. J Biol Chem 288:1489–1499. doi:10.1074/jbc.M112.419788
49. Martinez FO, Gordon S (2014) The M1 and M2 paradigm of macrophage activation: time for reassessment. F1000Prime Rep 6:13

50. Nguyen HX, Tidball JG (2002) Interactions between neutrophils and macrophages promote macrophage killing of rat muscle cells in vitro. J Physiol 547:125–132. doi:10.1113/jphysiol.2002.031450
51. Chen S-E, Gerken E, Zhang Y et al (2005) Role of TNF-α signaling in regeneration of cardiotoxin-injured muscle. Am J Physiol Cell Physiol 289:C1179–C1187. doi:10.1152/ajpcell.00062.2005
52. Langen RC, Van Der Velden JL, Schols AM et al (2004) Tumor necrosis factor-alpha inhibits myogenic differentiation through MyoD protein destabilization. FASEB J 18:227–237. doi:10.1096/fj.03-0251com
53. Arnold L, Henry A, Poron F et al (2007) Inflammatory monocytes recruited after skeletal muscle injury switch into antiinflammatory macrophages to support myogenesis. J Exp Med 204:1057–1069. doi:10.1084/jem.20070075
54. Ruffell D, Mourkioti F, Gambardella A et al (2009) A CREB-C/EBPβ cascade induces M2 macrophage-specific gene expression and promotes muscle injury repair. Proc Natl Acad Sci U S A 106:17475–17480. doi:10.1073/pnas.0908641106
55. Perdiguero E, Sousa-Victor P, Ruiz-Bonilla V et al (2011) p38/MKP-1-regulated AKT coordinates macrophage transitions and resolution of inflammation during tissue repair. J Cell Biol 195:307–322. doi:10.1083/jcb.201104053
56. Mounier R, Théret M, Arnold L et al (2013) AMPKα1 regulates macrophage skewing at the time of resolution of inflammation during skeletal muscle regeneration. Cell Metab 18:251–264. doi:10.1016/j.cmet.2013.06.017
57. Novak ML, Koh TJ (2013) Macrophage phenotypes during tissue repair. J Leukoc Biol 93:875–881. doi:10.1189/jlb.1012512
58. Villalta SA, Nguyen HX, Deng B et al (2008) Shifts in macrophage phenotypes and macrophage competition for arginine metabolism affect the severity of muscle pathology in muscular dystrophy. Hum Mol Genet 18:482–496. doi:10.1093/hmg/ddn376
59. Wermuth PJ, Jimenez SA (2015) The significance of macrophage polarization subtypes for animal models of tissue fibrosis and human fibrotic diseases. Clin Transl Med 4:2. doi:10.1186/s40169-015-0047-4
60. Forbes SJ, Rosenthal N (2014) Preparing the ground for tissue regeneration: from mechanism to therapy. Nat Med 20:857–869. doi:10.1038/nm.3653
61. Tonkin J, Temmerman L, Sampson RD et al (2015) Monocyte/macrophage-derived IGF-1 orchestrates murine skeletal muscle regeneration and modulates autocrine polarization. Mol Ther 23:1189–1200. doi:10.1038/mt.2015.66
62. Burzyn D, Kuswanto W, Kolodin D et al (2013) A special population of regulatory T cells potentiates muscle repair. Cell 155:1282–1295. doi:10.1016/j.cell.2013.10.054
63. Villalta SA, Rosenthal W, Martinez L et al (2014) Regulatory T cells suppress muscle inflammation and injury in muscular dystrophy. Sci Transl Med 6:10. doi:10.1126/scitranslmed.3009925
64. Rybalko V, Hsieh PL, Merscham-Banda M et al (2015) The development of macrophage-mediated cell therapy to improve skeletal muscle function after injury. PLoS One 10:1–19. doi:10.1371/journal.pone.0145550
65. Jetten N, Verbruggen S, Gijbels MJ et al (2014) Anti-inflammatory M2, but not pro-inflammatory M1 macrophages promote angiogenesis in vivo. Angiogenesis 17:109–118. doi:10.1007/s10456-013-9381-6
66. Kumar VA, Taylor NL, Shi S et al (2015) Self-assembling multidomain peptides tailor biological responses through biphasic release. Biomaterials 52:71–78. doi:10.1016/j.biomaterials.2015.01.079
67. Carlson BM (2003) Muscle regeneration in amphibians and mammals: passing the torch. Dev Dyn 226:167–181
68. Kragl M, Knapp D, Nacu E et al (2009) Cells keep a memory of their tissue origin during axolotl limb regeneration. Nature 460:60–65
69. Makanae A, Satoh A (2012) Early regulation of Axolotl limb regeneration. Anat Rec 295:1566–1574

70. Christensen RN, Tassava RA (2000) Apical epithelial cap morphology and fibronectin gene expression in regenerating axolotl limbs. Dev Dyn 217:216–224
71. Christensen RN, Weinstein M, Tassava RA (2002) Expression of fibroblast growth factors 4, 8, and 10 in limbs, flanks, and blastemas of Ambystoma. Dev Dyn 223:193–203
72. Godwin J, Kuraitis D, Rosenthal N (2014) Extracellular matrix considerations for scar-free repair and regeneration: insights from regenerative diversity among vertebrates. Int J Biochem Cell Biol 56:47–55. doi:10.1016/j.biocel.2014.10.011
73. Morrison JI, Loof S, He P et al (2006) Salamander limb regeneration involves the activation of a multipotent skeletal muscle satellite cell population. J Cell Biol 172:433–440
74. Tanaka EM, Gann AA, Gates PB et al (1997) Newt myotubes reenter the cell cycle by phosphorylation of the retinoblastoma protein. J Cell Biol 136:155–165
75. Sorokin L (2010) The impact of the extracellular matrix on inflammation. Nat Rev Immunol 10:712–723
76. Konttinen YT, Kaivosoja E, Stegaev V et al (2011) Extra-cellular matrix and tissue regeneration. Regener Med 22:21–80
77. Nambiar VV, Bhatt IY, Deshmukh PA et al (2008) Assessment of extracellular matrix remodeling during tail regeneration in the lizard *Hemidactylus flaviviridis*. J Endocrinol Reprod 12:67–72
78. Vinarsky V, Atkinson DL, Stevenson TJ et al (2005) Normal newt limb regeneration requires matrix metalloproteinase function. Dev Biol 279:86–98
79. Kovacs EJ (1991) Fibrogenic cytokines: the role of immune mediators in the development of scar tissue. Immunol Today 12:17–23
80. Moyer AL, Wagner KR (2011) Regeneration versus fibrosis in skeletal muscle. Curr Opin Rheumatol 23:568–573
81. Godwin JW, Pinto AR, Rosenthal NA (2013) Macrophages are required for adult salamander limb regeneration. Proc Natl Acad Sci U S A 110:9415–9420
82. Seifert AW, Monaghan JR, Voss SR, Maden M (2012) Skin regeneration in adult axolotls: a blueprint for scar-free healing in vertebrates. PLoS One 7, e32875
83. Echeverri K, Clarke JD, Tanaka EM (2001) In vivo imaging indicates muscle fiber dedifferentiation is a major contributor to the regenerating tail blastema. Dev Biol 236:151–164
84. Hay ED (1959) Electron microscopic observations of muscle dedifferentiation in regenerating *Ambystoma* limbs. Dev Biol 1:555–585
85. Lo DC, Allen F, Brockes JP (1993) Reversal of muscle differentiation during urodele limb regeneration. Proc Natl Acad Sci U S A 90:7230–7234
86. Duckmanton A, Kumar A, Chang YT et al (2005) A single cell analysis of myogenic dedifferentiation induced by small molecules. Chem Biol 12:1117–1126
87. Calve S, Simon HG (2011) High resolution three-dimensional imaging: evidence for cell cycle reentry in regenerating skeletal muscle. Dev Dyn 240:1233–1239
88. Sandoval-Guzmán T, Wang H, Khattak S et al (2014) Fundamental differences in dedifferentiation and stem cell recruitment during skeletal muscle regeneration in two salamander species. Cell Stem Cell 14(2):174–187
89. Echeverri K, Tanaka EM (2002) Mechanisms of muscle dedifferentiation during regeneration. Semin Cell Dev Biol 13:353–360
90. Kumar A, Velloso CP, Imokawa Y et al (2004) The regenerative plasticity of isolated urodele myofibers and its dependence on MSX1. PLoS Biol 2, E218
91. Pajcini KV, Corbel SY, Sage J et al (2010) Transient inactivation of Rb and ARF yields regenerative cells from postmitotic mammalian muscle. Cell Stem Cell 7:198–213
92. Kodaka Y, Tanaka K, Kitajima K et al (2016) LIM homeobox transcription factor Lhx2 inhibits skeletal muscle differentiation in part via transcriptional activation of Msx1 and Msx2. Exp Cell Res 331:309–319
93. Odelberg SJ, Kollhoff A, Keating MT (2000) Dedifferentiation of mammalian myotubes induced by msx1. Cell 103:1099–1109
94. Yang Z, Liu Q, Mannix RJ et al (2014) Mononuclear cells from dedifferentiation of mouse myotubes display remarkable regenerative capability. Stem Cells 32:2492–2501. doi:10.1002/stem.1742

95. Yilmaz A, Engeler R, Constantinescu S et al (2015) Ectopic expression of Msx2 in mammalian myotubes recapitulates aspects of amphibian muscle dedifferentiation. Stem Cell Res 15:542–553. doi:10.1016/j.scr.2015.09.012
96. Camarda G, Siepi F, Pajalunga D et al (2004) A pRb-independent mechanism preserves the postmitotic state in terminally differentiated skeletal muscle cells. J Cell Biol 167:417–423
97. Sacco A, Siepi F, Crescenzi M (2003) HPV E7 expression in skeletal muscle cells distinguishes initiation of the postmitotic state from its maintenance. Oncogene 22:4027–4034
98. Brookes S, Rowe J, Gutierrez Del Arroyo A et al (2004) Contribution of $p16^{INK4a}$ to replicative senescence of human fibroblasts. Exp Cell Res 298:549–559
99. Gilley J, Fried M (2001) One *INK4* gene and no ARF at the *Fugu* equivalent of the human *INK4A/ARF/INK4B* tumour suppressor locus. Oncogene 20:7447–7452
100. Kim SH, Mitchell M, Fujii H et al (2003) Absence of p16INK4a and truncation of ARF tumor suppressors in chickens. Proc Natl Acad Sci U S A 100:211–216
101. Calve S, Odelberg SJ, Simon HG (2010) A transitional extracellular matrix instructs cell behavior during muscle regeneration. Dev Biol 344:259–271
102. Calve S, Simon HG (2012) Biochemical and mechanical environment cooperatively regulate skeletal muscle regeneration. FASEB J 26:2538–2545
103. Morrison JI, Borg P, Simon A (2010) Plasticity and recovery of skeletal muscle satellite cells during limb regeneration. FASEB J 24:750–756
104. Grogg MW, Call MK, Okamoto M et al (2005) BMP inhibition-driven regulation of six-3 underlies induction of newt lens regeneration. Nature 438:858–862
105. Suetsugu-Maki R, Maki N, Nakamura K et al (2012) Lens regeneration in axolotl: new evidence of developmental plasticity. BMC Biol 10:103
106. King MW, Neff AW, Mescher AL (2012) The developing Xenopus limb as a model for studies on the balance between inflammation and regeneration. Anat Rec 295:1552–1562. doi: 10.1002/ar.22443
107. Flajnik MF, Hsu E, Kaufman JF et al (1987) Changes in the immune system during metamorphosis of Xenopus. Immunol Today 8:58–64
108. Robert J, Cohen N (1998) Evolution of immune surveillance and tumor immunity: studies in *Xenopus*. Immunol Rev 166:231–243
109. Grow M, Neff AW, Mescher AL et al (2006) Global analysis of gene expression in *Xenopus* hindlimbs during stage-dependent complete and incomplete regeneration. Dev Dyn 235:2667–2685
110. King MW, Neff AW, Mescher AL (2009) Proteomics analysis of regenerating amphibian limbs: changes during the onset of regeneration. Int J Dev Biol 295:955–969
111. Pearl EJ, Barker D, Day RC et al (2008) Identification of genes associated with regenerative success of *Xenopus laevis* hindlimbs. BMC Dev Biol 8:66
112. Fukazawa T, Naora Y, Kunieda T (2009) Suppression of the immune response potentiates tadpole tail regeneration during the refractory period. Development 136:2323–2327
113. Chen G, Robert J (2011) Antiviral immunity in amphibians. Viruses 3:2065–2086
114. Cotter JD, Storfer A, Page RB (2008) Transcriptional response of Mexican axolotls to *Ambystoma tigrinum* virus (ATV) infection. BMC Genomics 9:493
115. Kaufman J, Volk H, Wallny HJ (1995) A "minimal essential Mhc" and an "unrecognized Mhc": two extremes in selection for polymorphism. Immunol Rev 143:63–88

Chapter 4
Cartilage Healing, Repair, and Regeneration: Natural History to Current Therapies

Thomas P. Lozito, Yangzi Jiang, and Rocky S. Tuan

Abbreviations

ACI/ACT	Autologous chondrocyte implantation/transplantation
BMSC	Bone marrow stromal cell
BMP	Bone morphogenetic protein
Col	Collagen type
CR	Cartilage rod
CT	Cartilage tube
ECM	Extracellular matrix
ESC	Embryonic stem cells
FGF	Fibroblast growth factor
GAG	Glycosaminoglycan
IHH	Indian hedgehog
iPSC	Induced pluripotent stem cell
OA	Osteoarthritis
PG	Proteoglycan
lpr	Lymphoproliferative
MACI/MACT	Matrix-associated autologous chondrocyte implantation/transplantation
MMP	Matrix metalloproteinase
MSC	Mesenchymal stem cell
MRL	Murphy Roths Large
SHH	Sonic hedgehog

T.P. Lozito • Y. Jiang • R.S. Tuan (✉)
Department of Orthopaedic Surgery, Center for Cellular and Molecular Engineering, University of Pittsburgh School of Medicine, 450 Technology Drive, Room 221, Pittsburgh, PA 15219, USA
e-mail: tomlozito@gmail.com; yaj6@pitt.edu; rst13@pitt.edu

J. Wilson-Rawls, K. Kusumi (eds.), *Innovations in Molecular Mechanisms and Tissue Engineering*, Stem Cell Biology and Regenerative Medicine,
DOI 10.1007/978-3-319-44996-8_4

Sox	Sry-related high-mobility-group box
TGF	Transforming growth factor
VEGF	Vascular endothelial growth factor

4.1 Cartilage Structure and Function

Cartilage is a specialized tissue with several interesting characteristics that highlight a trade-off between function and healing. On the one hand, cartilage is an incredibly robust tissue, with the principal function of providing mechanical support, especially in weight-bearing circumstances. On the other hand, most cartilage exhibits almost a complete lack of intrinsic healing abilities once damaged. These two characteristics, mechanical durability and healing resistance, both stem from the unique structure of cartilage. Adult cartilage tissue is composed of over 90 % of extracellular matrix (ECM) and less than 10 % chondrocytes in total volume [1]. Thus cartilage is considered hypocellular, with few cartilage cells (chondrocytes) embedded in an abundant ECM. It is the molecular compositions of the cartilage ECM that define its mechanophysical properties: Proteoglycans (PGs) are responsible for the osmotic swelling and the elastic properties of the cartilage tissue. The most abundant cartilage PG, aggrecan, contains a core protein complexed with covalently bound glycosaminoglycan (GAG) side chains of chondroitin sulfate and keratin sulfate. Aggrecan further associates with hyaluronic acid filaments via link proteins. These PGs, which are negatively charged, attract cations and associated water molecules. The charged GAG side chains also repel one another, thereby trapping more water and causing the cartilage tissue to swell in the absence of physical load. In fact, the cartilage ECM contains 65–80 % water in wet weight [2]. Upon application of load, the hydrated GAG side chains allow cartilage to resist compression as water is forced from the tissue. When cyclically loaded, this ebb and flow of liquid through the cartilage tissue enables nutrient transport to chondrocytes [2]. Another important component of the cartilage ECM, the cartilage network, is responsible for the tensile strength of the cartilage matrix [1]. Collagen, the most abundant ECM component in the body, is a triple helical macromolecule with a cross-banded fibrillar structure that also acts as a meshwork that traps large PGs. The main collagen found in cartilage is collagen type II (Col2), but variations in the amounts of other collagen types and ECM components dictate the precise properties of the cartilage further classified as fibrocartilage, elastic cartilage, and hyaline cartilage [2]. Fibrocartilage is characterized by the inclusion of collagen type I (Col1) in the ECM and is found, for example, in the annulus fibrosus of intervertebral discs, the menisci, the pubic symphysis, and the temporomandibular joint. Elastic cartilage contains high amount of elastin and is found in the outer ear (auricular cartilage), the Eustachian tube, and the epiglottis. Hyaline cartilage matrix contains high amounts of Col2, chondroitin sulfate and hyaluronan, and is found on the ventral surfaces of ribs, in the larynx, trachea, and bronchi, and on the articular surfaces of bones (articular cartilage), where it is responsible for load bearing and shock absorption. Articular cartilage is the most clinically relevant form of cartilage as it is intimately involved in the pathogenesis of osteoarthritis (OA), and the last section of this chapter will focus specifically on articular cartilage healing.

While the high matrix-to-cell ratio of cartilage tissue underlies its mechanical properties, it also is responsible for its poor intrinsic healing capacities. In addition to being hypocellular, healthy adult cartilage is also avascular. Thus, injured cartilage has very few reserve chondrocytes available to synthesize new matrix. The chondrocytes that are present are trapped in their lacunae and embedded in dense cartilaginous matrix, making migration to wound sites difficult. Similarly, the lack of blood vessels also presents a barrier for stem cells from other parts of the body to reach the injured cartilage. Once cartilage tissue structure is compromised by a wound, the important nutrient transport environment begins to break down, causing loss of additional chondrocytes and cartilage tissue. Thus, rather than healing, even minor cartilage injuries can result in positive feedback scenarios in which large areas of cartilage are lost and do not regrow. Here we will examine special cases in the animal kingdom where cartilage does, in fact, naturally regenerate, as well as strategies for the therapeutic enhancement of cartilage healing.

4.2 Cartilage Formation During Embryonic Development and Adult Fracture Healing

Cartilage is initially formed in vertebrates during embryonic development of the skeletal system [3]. In fact, the early skeleton is entirely made up of cartilage, and cartilage cell sources vary with body location. For example, cartilage of the head is formed from the neural crest. Cartilage of the neck and trunk forms as part of the axial skeleton from the sclerotome of paraxial mesoderm, while cartilage of the tail skeleton originates from tail bud mesenchyme. Limb cartilage originates with the appendicular skeleton from lateral plate mesoderm. In the earliest stages of chondrogenesis, mesenchymal cells aggregate and condense in response to signaling molecules such as transforming growth factor-β (TGFβ), sonic hedgehog (SHH), and bone morphogenetic protein (BMP). Upon commitment to chondrogenesis, cells express the transcription factor Sox-9, which drives expression of cartilage-specific genes, including the matrix proteins Col2 and aggrecan. In vertebrates that undergo skeletal ossification, the cartilaginous skeleton acts as a template for the eventual replacement with bone, a process known as endochondral ossification. Chondrocytes cease proliferating and undergo hypertrophy. This critical milestone in the process of endochondral ossification is typified by characteristic changes in chondrocyte morphology, including dramatic increases in cell volume, and a defined gene expression profile. Hypertrophic chondrocytes begin secreting a unique matrix consisting of collagen type X and alkaline phosphatase, which initiates matrix calcification [4–6]. The hypertrophic chondrocytes also begin secreting the protease, matrix metalloproteinase-13 (MMP-13) [7–10], that breaks down cartilage matrix, and growth factors such as vascular endothelial growth factor (VEGF) [11], which induces blood vessels to sprout from the surrounding tissues. The hypertrophic chondrocytes then undergo apoptosis and are replaced by mesenchymal cells and pre-osteoblasts brought into the cartilage template via invading capillaries [12–15].

The remnant cartilage matrix is further cleared by invading osteoclasts and replaced with bone matrix as mesenchymal cells differentiate into osteoblasts. Endochondral ossification concludes when the cartilage template is replaced by bone. Not all embryonic cartilage is replaced by bone, however, and the permanent cartilage that persists following embryonic development make up the fibrocartilage, elastic cartilage, and hyaline cartilage of the adult organism.

Interestingly, many of the same milestones observed in embryonic cartilage and skeletal development are also seen in adult vertebrate fracture healing [16]. Furthermore, the primary morphogenetic pathways that are active during embryonic skeletal development are also expressed in fracture calluses, and a comparison of the transcriptomes has revealed that genes that control appendicular limb development also show increased expression during fracture healing [17]. Fracture healing begins with an initial anabolic phase characterized by an increase in tissue volume related to the de novo recruitment and differentiation of stem cells that form skeletal and vascular tissues. The tissue between broken bones at the fracture site swells as hematomas form. The adjacent periostium also swells, and periosteal stem/progenitor cells proliferate into the fracture. These cells undergo chondrogenesis, forming the cartilage callus. Concurrent with cartilage tissue development, cells that will form the nascent blood vessels that supply the new bone are recruited and differentiate in the surrounding muscle sheath. As chondrocyte differentiation progresses through hypertrophy, the cartilage extracellular matrix undergoes mineralization and the anabolic phase of fracture repair terminates with chondrocyte apoptosis. Just as in endochondral ossification, blood vessels invade in response to VEGF signals, bringing pre-osteoblasts that replace cartilage tissue with bone. The anabolic phase is followed by a prolonged phase in which catabolic activities predominate as the callus is resorbed and remodeled to the bone's original cortical structure. The recapitulation of these ontological processes is believed to make fracture healing one of the few postnatal processes that is truly regenerative, restoring the damaged skeletal organ to its pre-injury cellular composition, structure and biomechanical function [16]. As discussed in the following section, certain non-mammalian organisms are capable of even more impressive feats of regeneration.

4.3 Cartilage Regeneration During Limb/Tail Regeneration

Several remarkable organisms are able to regenerate amputated limbs and/or tails. In doing so, the tissues of the lost appendage are replaced, including cartilage. In fact, cartilage is the default skeletal tissue for appendage regeneration, and, in these special cases, the regenerated cartilage does not ossify for the lifetime of the regenerate. These feats of regeneration are achieved through processes that meld embryonic development with adult wound healing, and what we learn from them may offer clues for improving mammalian regeneration.

Urodeles (salamanders and newts) and *Xenopus* frogs are able to regenerate limbs as adults (Table 4.1). While urodeles are able to regenerate both front and back limbs, frogs are able to regenerate front limbs only. Urodeles retain non-ossified,

Table 4.1 Comparison of vertebrate cartilage regeneration and healing abilities

Organism		Adult skeleton	Limb regeneration	Tail regeneration	Full thickness articular defect	Partial thickness articular defect	Ear hole closure
Amphibia	Urodele	Cartilaginous	Yes [18, 19]	Yes [20–22]	Yes [23]	?	NA
	Frog	Osseous	Yes [18]	NA	?	?	NA
Reptilia	Lizard	Osseous	No	Yes [24–26]	Yes [27]	?	NA
Mammalia	*Acomys* mice	Osseous	No	No	?	?	Yes [28]
	MRL mice	Osseous	No	No	Yes [29]	No [29]	Yes [30]
	Wild type mice	Osseous	No	No	No	No	No
	Humans	Osseous	No	No	No	No	No

cartilaginous skeletons into adulthood and are able to regenerate fully formed limbs (Fig. 4.1a), with all the cartilaginous skeletal elements of the originals (Fig. 4.1c). Regenerated urodele limbs also recreate the musculature of the amputated arms/legs. Frogs, which do fully develop and exhibit ossified skeleton as adults, regenerate cartilage spikes rather than limbs following amputation (Fig. 4.1b). Cartilage spikes are continuous with the radio-ulna bone of the original limb, and no other skeletal elements are formed, and very little muscle is regenerated (Fig. 4.1d). These differences between urodele and frog limb regeneration are remarkable given that both processes begin very similarly. Following limb loss in both animal groups, limb stump tissues contract and wound epidermis forms to seal the stump. Stump tissues are broken down by secreted proteases, releasing cells into the stump. These cells migrate and proliferate, forming the blastema, the classic indicator of regeneration, and blastema cells reform the majority of tissues of the replacement limb. In frogs, regenerated spike cartilage does not originate from blastema cells. Instead, cartilage spikes originate from severed bones of amputated limbs and are formed similar to cartilage calluses during fracture repair [18]. How this callus-like accumulation of cartilage extends into the spikes of regenerated frog limbs is not known currently, but may provide clues for healing bone fracture gaps that exceed critical size defect lengths. In urodeles, regenerated limb cartilage does originate from blastema cells (Fig. 4.2) [19]. Interestingly, the blastema cells that differentiate into cartilage are derived from both the dermis and cartilage of the original limb stump, but not from muscle. This restriction in differentiation stems from the fact that the urodele limb blastema is a heterogeneous collection of restricted progenitor cells that do not cross developmental origins as they reform lost tissues. For example, both dermis and skeletal tissues originate from the lateral plate mesoderm, and blastema cells originating from either of these tissues are able to differentiate into cartilage, but not muscle, which originates from presomitic mesoderm. Similarly, limb blastema cells derived from muscle do not differentiate into dermis or cartilage.

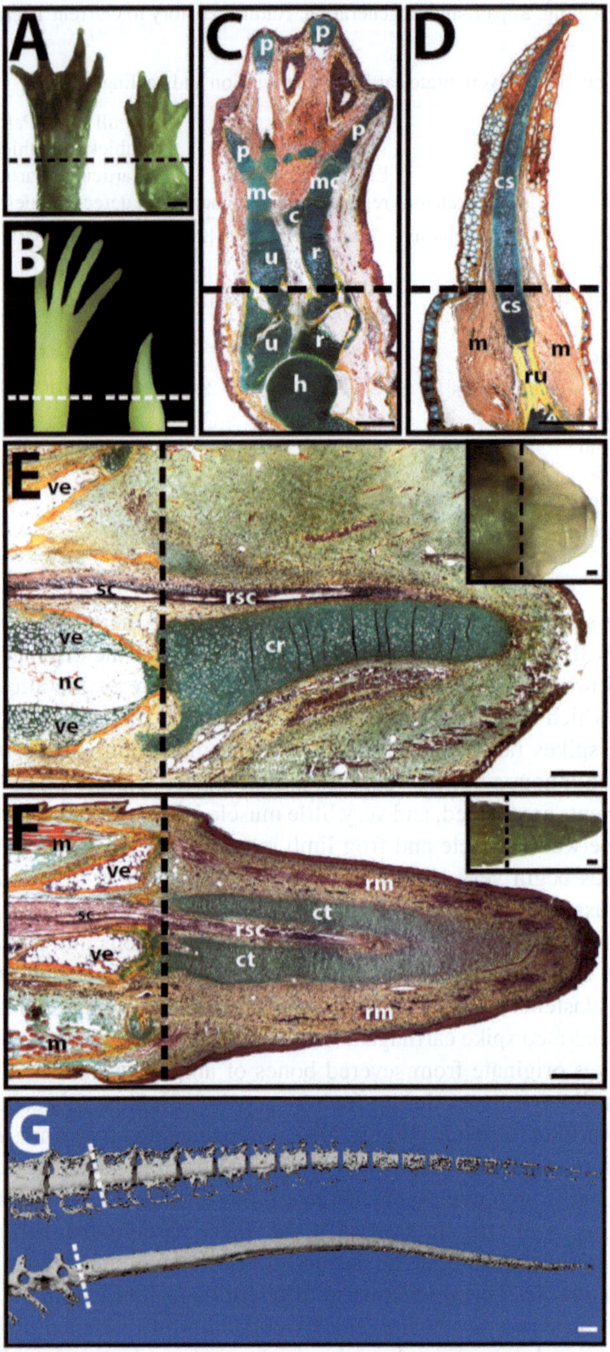

Fig. 4.1 Examples of limb and tail regeneration in amphibians and lizards. (**a**, **b**) Morphological comparison of (**a**) salamander (*Ambystoma mexicanum*) and (**b**) frog (*Xenopus laevis*) forelimbs before (*left*) and 8 weeks after (*right*) amputation. Salamanders regenerate new limbs, while frogs regenerate cartilage spikes. (**c**, **d**) Histological analysis (pentachrome) of regenerated (**c**) salamander and (**d**) frog limbs. Salamanders regenerate all the skeletal elements of the upper arm and hand,

The inverse relationship between complexity and regeneration fidelity and the preference for producing cartilage noted for limb regeneration are also observed in tail regeneration. Urodeles and lizards regenerate tails (Table 4.1) [24–26, 32], and both regenerated tail skeletons are almost completely cartilaginous (Fig. 4.1e, f). Salamanders regenerate cartilage rods (CR) ventral to regenerated spinal cords (Fig. 4.1e), while lizards regenerate cartilage tubes (CT) that enclose regenerated spinal cords (Fig. 4.1f). However, regenerated tails of the comparatively primitive salamander segment and develop neural and hemal arches, and mature regenerated salamander tails are almost perfect copies of originals (Fig. 4.1g). The more complex lizards, on the other hand, regrow imperfect regenerated tails, and lizard cartilage tubes never segment and are easily distinguishable from original tail skeletons (Fig. 4.1g). Also unlike salamander cartilage regeneration, a portion of the regenerated lizard cartilage ossifies [24]. The most proximal region of the CT in contact with the original tail skeleton undergoes endochondral ossification in a process similar to what is observed during fracture healing. Proximal CT chondrocytes undergo hypertrophy and are replaced by bone. This proximal ossification event is not observed in the urodele CR, and may reflect the differences in ossification states between adult urodele and lizard skeletons. Interestingly, the perichondrium of the distal lizard CT calcifies without undergoing ossification, while the CT interior remains cartilaginous for the lifetime of the regenerate. Like bone periosteum, the lizard CT perichondrium harbors a stem/progenitor cell population that forms additional cartilage in response to stimulation with TGFβ [24]. Like urodele regenerated cartilage, cartilage formed from lizard CT perichondrium cells does not undergo hypertrophy and endochondral ossification. These observations also indicate a link between original and regenerated cartilage ossification: cartilage formed by cells derived from ossified tissues undergo hypertrophy and ossification, while cartilage derived from cartilaginous tissue elements do not. This topic becomes important during discussion of cell therapies for cartilage healing in humans, which are plagued by unwanted cartilage hypertrophy and ossification.

Tail regeneration also provides an interesting contrast to limb regeneration in terms of cell identity. As with limb regeneration, urodele and lizard tail generation begins with blastemas. Unlike limb blastema cells, whose differentiation is lineage restricted by developmental origin (i.e., mesoderm vs ectoderm) [19], tail blastema

Fig. 4.1 (continued) while frogs regenerate a single cartilage spike. (**e**, **f**) Histological (pentachrome) and (**e**, **f** Insets) morphological analysis of (**e**) salamander tail 5-weeks post amputation and (**f**) lizard (*Anolis carolinensis*) tail 2 weeks post-amputation. (**g**) Salamander (*top*) and lizard (*bottom*) tails 10 weeks after amputation analyzed by micro-computed tomography. Pentachrome stains cartilage *green*, bone *orange*, muscle *red*, and spinal cord and epidermis *purple*. *Dashed lines* denote amputation planes. *c* carpal, *cr* cartilage rod, *cs* cartilage spike, *ct* cartilage tube, *h* humerus, *m* muscle, *mc* metacarpal, *nc* notochord, *p* phalanges, *r* radius, *rm* regenerated muscle, *rsc* regenerated spinal cord, *ru* radio-ulna, *sc* spinal cord, *u* ulna, *ve* vertebra. Bar = 1 mm. Figure adapted from [31]

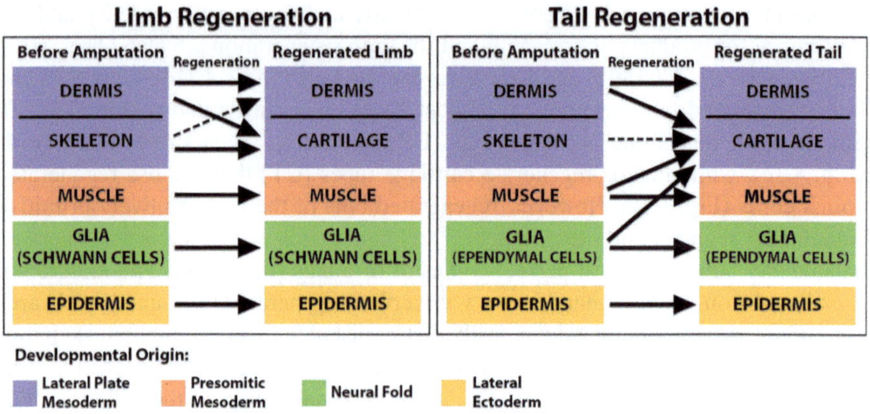

Fig. 4.2 Summary of blastema cell differentiation restrictions during salamander limb and tail regeneration. Figure adapted from [19]

cells are able to transition between developmental lineages during differentiation (Fig. 4.2). For example, regenerated tail cartilage is formed from blastema cells that have originated from muscle [20], dermis, or even spinal cord (ependyma) [21]. In fact, tail skeletal tissue contributes only minimally to regenerating tail blastemas and, hence, regenerated cartilage. The reasons for these differences between limb and tail blastema origins are not currently understood, but they may reflect differences involving both development and healing in the appendicular versus axial skeletons.

In summary, appendage regeneration is depended on the formation of wound epithelia and blastemas or blastema-like structures. This encapsulation of proliferating cells by un-differentiated, embryonic-like epithelial tissue provides the necessary environment for tissue differentiation and extension and avoids scar formation. While lizards, urodeles, and frogs provide perhaps the best examples of these structures and the regenerative process, certain mammals are also capable of approximating these healing responses.

4.4 Cartilage Healing and Regeneration in Non-Human Mammals

As a group, mammals exhibit much reduced regenerative abilities compared to amphibians and lizards. For example, no mammal is capable of limb or tail regeneration as adults. While some rodent species, such as African spiny mice (*Acomys*) and South American spiny rats (*Proechimys*) shed tails as strategies for escaping predators (caudal autotomy), lost tails are not regenerated [33, 34]. Perhaps the most impressive naturally-occurring examples of adult regeneration among mammals are observed in species capable of skin autotomy. For example, the skin of

Acomys mice is mechanically weak and easily tears and sloughs off [28]. This makes *Acomys* mice difficult for predators to grab and hold onto, allowing the mice to escape, but often results in large open wounds and skin loss. *Acomys* mice, but not house mice (*Mus*), are able to heal these types of skin wounds quickly and without scarring [28]. In the lab, *Acomys* mice are capable of healing ear hole punches, including auricular cartilage (Table 4.1) [28]. Interestingly, these types of mice appear to generate blastema-like structures during healing, as evidenced by wound epidermis that bears striking similarities to those formed during appendage regeneration in urodeles and lizards. Whether the *Acomys* blastema follows the same rules in cell fate and differentiation remains to be determined.

While *Acomys* mice and *Proechimys* rats may represent the best examples of "natural" cartilage regeneration among mammals, certain mouse strains exhibit enhanced regenerative abilities following selective breeding over many generations. The so-called "super healing" mouse strains are able to heal a number of tissues better than wild type mice. Collectively known as the Murphy Roths Large (MRL) mice, this groups includes the MRL/MpJ, Murphy Roths Large/lymphoproliferative (lpr) mouse strain (MRL/MpJ-*Fas*^{lpr}/J) MRL/MpJ-*Fas*^{lpr}/J, and Large strains [35]. Like *Acomys* mice, MRL mice are able to heal ear hole punches and regenerate auricular cartilage (Table 4.1) [30]. In addition, MRL mice form a type of wound epidermis faster than other strains, and appear to form blastema-like accumulations of mesenchymal cells in response to certain types of injuries. It is interesting that neither *Acomys* mice nor MRL strains are able to regenerate limbs, tails, or digit tips as adults [36].

While direct comparisons between *Acomys* mice and MRL strains have yet to be made, based on their similar abilities to form blastema-like structures and heal hole punch injuries it is possible that similar healing mechanisms are at work in both animals. Unfortunately, the exact underlying mechanisms responsible for the enhanced healing abilities of MRL mice have proven difficult to specify. The "super healer" phenotype appears to depend most heavily on the inclusion of the Large strain identity, which includes autoimmune anomalies in addition to enhanced healing. For example, the MRL/MpJ-*Fas*^{lpr}/J strain was established through selective interbreeding of the B6 (0.3 %), C3H (12.1 %), AKR (12.6 %), and Large (75 %) strains [35]. These mice are prone to autoimmune disorders, and these phenotypes were attributed to a mutant *Fas* gene, which arose spontaneously at generation F12 during selective breeding. However, the link between mutant *Fas* and healing is confounded by the fact that the MRL/MpJ mice, which have the wild type *Fas* gene and were maintained as a control strain for the MRL/MpJ-*Fas*^{lpr}/J mice, also exhibit enhanced healing. Still, since all 3 MRL mouse strains exhibit autoimmune phenotypes, it is natural to suppose a link between regenerative ability and immunity dysfunction. However, a multi-strain wound healing survey offers evidence that they are not genetically linked. It should be noted, however, that mutations in the cell cycle checkpoint gene p21 cause yet another autoimmune disorder similar to lupus, but also enhanced healing phenotypes [37]. Obviously, additional research is needed to work out the mechanisms behind the "super healing" phenotypes (see comprehensive review by Heydeman [35]).

4.5 Intrinsic Articular Cartilage Repair

While cartilage regeneration activities in response to appendage amputation and ear hole punch injuries are informative in assessing the healing limits of non-human animals, they are admittedly not readily relatable to human cartilage injuries, which predominantly affect the articular cartilage of limb joints. As previously mentioned, highly specialized hyaline cartilage lines the articular surfaces of long bones. Articular cartilage itself is divided into four zones based on chondrocyte morphology, matrix composition and distribution: (1) superficial zone (tangential layer), consisting of two to three layers of small, flattened chondrocytes arranged parallel to the surface; (2) middle or transitional zone, where the chondrocytes are spherical; (3) deep or radial zone, consisting of large chondrocytes that form columns perpendicular to the surface; and (4) calcified zone, where hypertrophic chondrocytes are embedded in the calcified matrix, which is connected to the subchondral bone (Fig. 4.3). Differences in the ECM are seen within the hierarchical structure of articular cartilage. In the surface zone, chondrocytes produce proteoglycans that reduce friction (i.e., lubricin), protect chondrocytes and cartilage surfaces, and inhibit synovial cell overgrowth [39, 40]. In the middle zone, the ECM includes Col2, aggrecan, and other proteins. Collagen type X and alkaline phosphatase are found in the deep zone and calcified zone, indicating chondrocyte hypotrophy and the calcified matrix environment (Fig. 4.3). Collagen fibrils are oriented mostly parallel to the surface in the superficial zone, obliquely in the middle zone, and perpendicular to the joint surface in the deep zone, which is suited to load transmission (Fig. 4.3) [38].

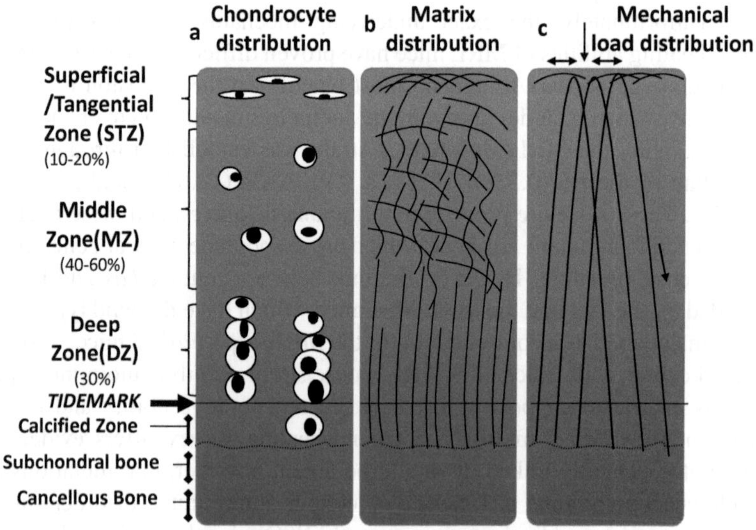

Fig. 4.3 Schematic view of normal articular cartilage highlighting (**a**) cell distribution, (**b**) matrix distribution, and (**c**) collagen fibril orientation. Figure modified from [38]

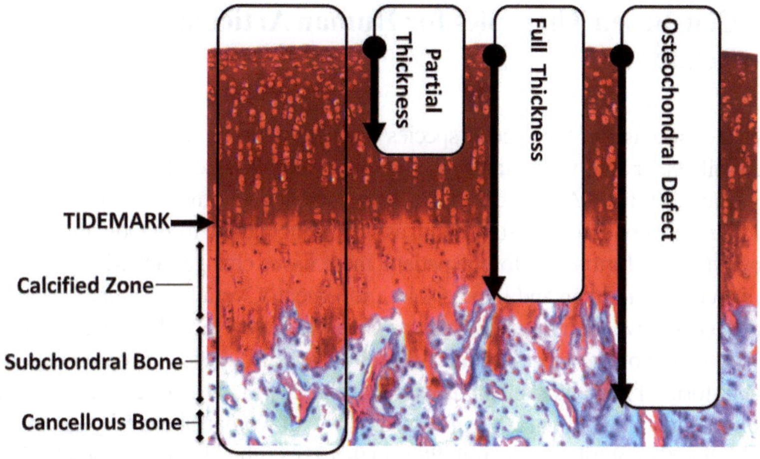

Fig. 4.4 Three different classes of cartilage injury dependent on the depth of defect (rabbit knee joint cartilage as background)

In terms of tissue repair ability, cartilage injury is classified on the basis of the depth of defect (Fig. 4.4), and tissue remodeling response differs depending on the type and size of the defect. Partial cartilage defects are limited to the superficial-to-middle zones and do not involve damage to subchondral bone. Full thickness cartilage defects penetrate down to the bone, and are in fact more prone to heal than partial thickness defects if the osteochondral junction is also damaged. In these cases, where full thickness defects penetrate into the bone marrow, bone marrow stromal cells (BMSCs) flow into the lesion site to form a stem-cell rich fibrin clot and stimulate intrinsic repair. While humans are unable to heal both partial and full thickness defects, some of the species discussed above do manage at least some level of articular cartilage healing. MRL mice are able to heal full thickness defects up to 0.5 mm^2 in depth, and the regenerated cartilage is robust hyaline cartilage (Table 4.1) [29]. However, even the "super healing" mouse strains are unable to heal partial thickness defects (Table 4.1). Articular cartilage healing has yet to be studied in *Acomys* mice and p21 knockout mice, and such experiments would provide interesting context for the results involving MRL mice. Among non-mammalian animals, lizards are able to regenerate entire articular cartilage surfaces (Table 4.1) [27], but new cartilage tissue appears to undergo hypertrophy and is probably more similar to a fracture cartilage callus than true regenerated articular cartilage [41]. It is currently not known if lizards or other reptiles can heal partial or full thickness cartilage defects. Salamanders can regenerate full thickness cartilage defects that cover approximately 50 % of the joint [23], but it is not yet known if urodeles can regenerate partial thickness defects (Table 4.1). In any case, these experiments not only provide evidence that healing large articular cartilage defects *is* possible, but they also provide hope that articular cartilage healing may be achieved by humans with the correct therapies.

4.6 Cell-Based Therapies for Human Articular Cartilage Repair

Unlike the cartilage of the special species discussed above, adult human cartilage has limited self-repair ability, and damage to articular cartilage leads directly to the pathogenesis of osteoarthritis (OA). For example, progressive loss of articular cartilage leads to an increase in subchondral bone formation, as well as new bone formation at joint margins (osteophytes). Unfortunately for the patient, these tissue changes underlie clinical symptoms including joint pain and limited joint movement. Overall, these pathologies manifest as degenerative joint diseases, such as OA, which severely affect the quality of life [42]. OA is one of the most common causes of mobility loss and represents the most prevalent form of musculoskeletal disease worldwide [43, 44]. For example, OA affects 27 million Americans, about 60 % of men and 70 % of women above 65 years of age [45, 46], and directly contributes to disabilities in 9–10 % of the U.S. population [47].

As mentioned above, humans do not spontaneously heal partial or full thickness cartilage defects, OA progresses until the entire affected joint needs to be either fused or replaced. However, there is evidence of incomplete healing in small and deep defects. Osteochondral defects do exhibit limited reparative capacity, and, in clinical practice, this intrinsic reparative property is exploited in the microfracture technique, which involves surgical drilling to the subchondral bone region to treat small size cartilage defects (usually 0.5–2 cm^2) [2]. However, the cartilage formed in response to subchondral microfracture consists mainly of fibrocartilage rather than the original hyaline cartilage, and the therapeutic benefits generally last only 2–5 years [48–50]. For larger defects that require more extensive healing, tissue transplantation such as osteochondral auto/allograft (mosaicplasty) has been used; however, tissue source and compatibility present potential complications. Most of the current approaches to treat articular cartilage injuries, therefore, have focused on stimulating intrinsic regeneration and/or replacing diseased or lost tissue. These therapeutic approaches are collectively known as tissue engineering and regenerative medicine, an area that has been developing rapidly since the 1970s. Termed the "next evolution of medical treatments" by the U.S. Department of Health and Human Services, regenerative medicine aims to replace or regenerate human cells, tissues and organs to restore or establish normal function [51]. The basic principle involves the application of cells, biomaterial scaffolds, and signaling molecules to promote endogenous regenerative capacity and/or the replacement of whole tissues with engineered constructs in vitro [52]. Regenerative medicine approaches for healing articular cartilage injuries offer promise for preventing OA.

4.6.1 Autologous Chondrocyte-Based Therapies for Cartilage Defects

The concept of autologous implantations to treat cartilage defects began with studies by O'Driscoll and co-workers, who used periosteal grafting to treat rabbit chondral defects [53]. Further refinement by Grande and Peterson included the

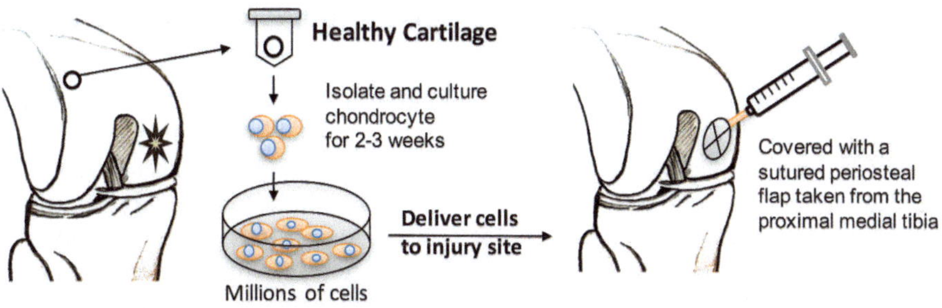

Fig. 4.5 Schematic of autologous chondrocyte implantation

use of cultured autologous chondrocytes [54, 55]. Autologous Chondrocyte Implantation/Transplantation (ACI/ACT) was first applied clinically to treat full-thickness chondral defects in knees by Brittberg et al. [56]. Briefly, small amounts of healthy cartilage were harvested from non-load bearing areas under arthroscopy, and the isolated chondrocytes were expanded in vitro for up to 6 weeks. The cultured cells were then injected into the cartilage defect and sealed with a sutured periosteal flap taken from the proximal medial tibia (Fig. 4.5). The overall 0–5 year therapeutic efficacy was generally 70–90 %, as evidenced by relief of symptoms and improvement of joint function [57]. In a 10–20 year (mean 12.8 year) follow-up study, 74 % of the 224 patients that underwent ACI treatment reported their status as good or better than before surgery [58]. ACI/ACT have also been reported to be effective in treating larger cartilage defects [59], with therapeutic benefits lasting longer than those of microfracture marrow-stimulation techniques [60]. Therefore, ACI provides the possibility of regenerating cartilage tissues and restoring normal joint function, criteria which meet the basic clinical definition for functional cartilage repair.

To eliminate the need for secondary surgery sites and to reduce the complexity of the ACI/ACT procedure, biomaterials have been adopted in the next generations of ACI/ACT. Standard procedure of ACI/ACT involves surgical preparation of the defect(s), periosteal harvesting, suturing of periosteum over defect(s), application of fibrin glue sealant, and implantation of chondrocytes with the risks of possible cell leakage from the application sites as well as uneven cell distributions. Furthermore, the harvesting of periosteum increases the operation time and requires a larger surgical exposure field [61]. To address these shortcomings, "second generation" ACI uses biomaterials (e.g., collagen type I/type III membranes) instead of periosteum grafts, thereby reducing open injury sites and shortening operation time. More recently, third generation, or "all in one" grafts, have been developed that make use of combinations of cells and biomaterials, which are delivered directly to defects without either periosteal covers or suture fixation. This technique is referred to as matrix-associated autologous chondrocyte implantation (MACI). Currently, the most commonly used biomaterials in MACI involve natural ECM materials such as collagen and hyaluronan [62], and there is active, ongoing research to develop more optimal biomaterials [62, 63].

4.6.2 Stem Cell-Based Therapies for Cartilage Regeneration

Despite the promise of ACI and MACI, limitations remain, and current research is aimed at improving therapeutic effectiveness and availability. For example, ACI and MACI are limited by the availability of harvested cell number and quality. In clinical application, chondrocytes directly derived from healthy hyaline cartilage are considered the most appropriate for transplantation [64]. Unfortunately, the numbers of chondrocytes suitable for harvest are very limited. For example, patients in need of ACI often have experienced extensive cartilage degeneration and loss; in addition, chondrocytes exhibit only limited life span as differentiated cells during culture expansion before cell quality irreversibly suffers. To address the shortage of suitable cell populations, stem cells that may serve as chondroprogenitors are under investigation as new candidate cell sources to replace native chondrocytes for cartilage repair.

Mesenchymal stem cells (MSCs) are the most promising therapeutic cells for cartilage regeneration research, owing to their self-renewal ability, chondrogenic potential, and anti-inflammatory activity [65]. Clinical application of bone marrow-derived MSCs has been reported by several groups [66–68], and a 2 year follow up cohort study showed comparable efficacy of MSCs and native chondrocytes for use in ACI [69]. However, longer term studies are already needed. One of the most important and interesting aspects of using MSCs in ACI is the dependency of MSC chondrogenic potential on cell source since, ultimately, the clinical outcome depends on the ability of the stem cells to form cartilage. A summary of studies evaluating the use of MSCs from various tissue sources in treating ACI in animal studies is presented in Table 4.2. This comparison indicates that bone marrow-derived MSCs produce more hyaline-like cartilage matrix and promote higher functional recovery than MSCs isolated from periosteum, synovium, adipose tissue, and muscle [70], which tend to undergo fibrocartilage differentiation [70, 95]. MSCs isolated from tissues other than bone marrow do offer certain advantages, however. For example, adipose-derived MSCs are easy to obtain, and adipose tissue contains 100-times greater numbers of stem cells per volume than bone marrow aspirates [96]. Unfortunately the chondrogenic potential of adipose-derived MSCs is lower compared to bone marrow MSCs [97], suggesting that more research needs to be done to improve the chondrogenic differentiation of these cells.

Furthermore, given their expanded levels of differentiation potencies, both embryonic stem cells (ESCs) and induced pluripotent stem cells (iPSCs) have the potential for chondrogenesis [94, 98] with the additional options of founding patient-specific cell lines with high self-renewal potential, these cells may be the ideal candidates for cartilage regenerative medicine. Indeed, animal studies have already been conducted [91–93, 99] (Table 4.2). However, several complications have yet to be overcome. For example, not all of the transplanted cells contribute to hyaline cartilage regeneration [93], and not all cell lines differentiate into the target tissue safely [99]. Thus, before ESC and iPSC cells are used in a clinical setting, the topics of differentiation efficiency and tumor formation must be solved.

Table 4.2 Human stem cells for cartilage regeneration

Cell sources		Animal model	General outcome
MSCs	Bone marrow	Rabbit [70–77] Pig [78] Goat [79] Rat [80, 81]	• Increased tissue formation and reduction in degenerated cartilage [71, 72] • Histological score improvement [71, 76] • Repaired cartilage was hyaline-like [70, 73, 75, 77, 80] • Restoration of mechanical properties [78] • Cartilage specific markers expression and cartilage formation, forming hyaline cartilaginous tissue [80]
	Adipose	Rat [82] Rabbit [70, 71, 83]	• Cells differentiated into functional chondrocytes that secreted cartilaginous matrix [70, 71, 82, 83] • Less repair than bone marrow-derived MSC [70]
	Synovium	Rabbit [71, 84–87] Minipig [88]	• High histological score improvement [71, 84–86] • Enhanced cartilage matrix production [87, 88] • Integrated with surrounding native cartilage [87]
	Periosteum	Rabbit [70]	• Increased histological grading [70]
	Muscle	Rat [89, 90] Rabbit [70, 71]	• Enhanced histological scores and ECM deposition [89, 90] • Less repair than BMSCs [70, 71]
	Umbilical cord	Rabbit [75]	• Enhanced histological score than no-cell control but lower than BMSC [75]
ESCs	Embryo	Rat [91, 92] Sheep [93]	• Produce cartilage, resulting in repair of defects without forming any teratomas [91] • Formed neocartilage layer with good surface regularity and complete integration [92] • Promoted better organization and tissue bulk, but no effect on histological evaluation [93]
iPSCs	Cells by reprogramming	Nude mice [94]	• Some cell lines formed tumors, others induced cell lines generated cartilage-like tissue, but others formed tumors [94]

4.6.3 *Future Studies to Enhance Stem Cell-Based Cartilage Regeneration*

Both the ACI and MACI techniques for cartilage repair are well-established examples of tissue engineering/regenerative medicine and represent the current best solutions for cartilage injury. However, neither ACI nor MACI can completely regenerate hyaline cartilage for large defects, and there is significant need for further improvements. An important area of current research involves the optimization of differentiation signals and environments for producing robust hyaline cartilage. Clues from embryonic cartilage developmental processes, as well as from cartilage regeneration in non-human animals, could shed light on such studies.

For example, to improve stem cell differentiation efficacy and maintain chondrocyte phenotype, signaling factors such as TGF-βs are required. However, a growing amount of evidence has indicated that treatment with single signaling factors is insufficient for initiating maximal stem cell chondrogenesis and phenotype maintenance. Thus, knowledge gained on embryonic skeletal system development/nonhuman cartilage regeneration should be used a guide.

Embryonic chondrogenesis begins with mesenchymal cell recruitment, proliferation and condensation. Cell condensations are initiated by several growth factors, including TGF-β, FGF, Wnt, and BMPs, acting in concert [3, 100, 101]. Afterwards, several matrix molecules, including fibronectin, hyaluronan and collagens, interact with the cell surface receptors to initiate the transition from chondroprogenitor to chondrocytes [100, 102, 103] and regulation of the chondrogenesis-specific transcription factor Sox-9 [104]. In an example of recreating multi-step differentiation schemes in vitro, ESCs/iPSCs were treated with two-step differentiation strategies. First, ESCs/iPSCs were differentiated into multipotent states (ESC-MSC or iPSC-MSC), which were then differentiated towards the chondrogenic linage [105]. These strategies offer promise for creating significant amounts of healthy cartilage, but additional work is required to fine-tune the differentiation signals. For human MSCs, TGFβ2 and TGFβ3 were shown to be more active than TGFβ1 in promoting chondrogenesis [106]. Interestingly, the effect of TGFβ3 stimulation is enhanced if the growth factor is applied during the initial phase of the culture period and then withdrawn [107, 108]. Adding to the complexity, the effects of growth factor treatments varies with MSC tissue source. For example, BMP6 in addition to TGFβs is required by adipose-derived stem cells for efficient stimulation of chondrogenesis [109, 110]. Again broadening our discussion to non-human animals, TGFβs and Indian hedgehog (Ihh) regulate cartilage formation and maturation during lizard tail regeneration [24]. TGFβ1 and TGFβ3 induce cartilage formation in lizard CT perichondral cells, which express the MSC markers CD90 and CD66, and the CT perichondrium calcifies in response to Ihh. Inhibiting hedgehog signaling in the regenerating lizard tail suppresses cartilage maturation, which may provide clues for preventing similar maturation in cartilage derived from progenitor cells in other species, including human MSCs. Indeed, considering the complex mixture of factors involved in embryonic skeletogenesis and appendage regeneration in vivo, we may surmise that a similarly complex, multifactorial biochemical environment will be required for effective long-term cartilage engineering.

4.7 Conclusion

Cartilage is a tissue that most animals, including humans, are unable to repair. In this chapter we have summarized the cartilage healing abilities of the few species which are able to regenerate cartilage. We have also described the current approaches in therapeutic enhancement of cartilage repair in humans. It is noteworthy that cartilage therapies may be adapted to mimic the pattern and sequence of biological events seen in naturally regenerative tissues. For example, the use of autologous stem cells to augment the resident progenitor cell population represents a strategy that echoes the role of the blastema in appendage regeneration. As future research works out the intricacies of cell differentiation and signaling, similar advancements will help in closing gaps in wound healing capabilities.

Acknowledgments Supported by the Commonwealth of Pennsylvania Department of Health (SAP 4100050913), US Department of Defense (W81XWH-14-2-000, OR130296, W81XWH-15-0104), Arthur J. Rooney, Sr. Endowment, and NIH (R01 GM115444).

References

1. Gentili C, Cancedda R (2009) Cartilage and bone extracellular matrix. Curr Pharm Des 15:1334–1348
2. Williams RJ (2007) Cartilage repair strategies. Humana Press, Totowa
3. DeLise AM, Fischer L, Tuan RS (2000) Cellular interactions and signaling in cartilage development. Osteoarthr Cartil 8:309–334
4. Anderson HC, Sipe JB, Hessle L, Dhanyamraju R, Atti E, Camacho NP, Millan JL (2004) Impaired calcification around matrix vesicles of growth plate and bone in alkaline phosphatase-deficient mice. Am J Pathol 164:841–847
5. Fedde KN, Blair L, Silverstein J, Coburn SP, Ryan LM, Weinstein RS, Waymire K, Narisawa S, Millan JL, MacGregor GR, Whyte MP (1999) Alkaline phosphatase knock-out mice recapitulate the metabolic and skeletal defects of infantile hypophosphatasia. J Bone Miner Res 14:2015–2026
6. van der Eerden BC, Karperien M, Wit JM (2003) Systemic and local regulation of the growth plate. Endocr Rev 24:782–801
7. Tchetina EV, Kobayashi M, Yasuda T, Meijers T, Pidoux I, Poole AR (2007) Chondrocyte hypertrophy can be induced by a cryptic sequence of type II collagen and is accompanied by the induction of MMP-13 and collagenase activity: implications for development and arthritis. Matrix Biol 26:247–258
8. Cawston TE, Wilson AJ (2006) Understanding the role of tissue degrading enzymes and their inhibitors in development and disease. Best Pract Res Clin Rheumatol 20:983–1002
9. Inada M, Wang Y, Byrne MH, Rahman MU, Miyaura C, Lopez-Otin C, Krane SM (2004) Critical roles for collagenase-3 (Mmp13) in development of growth plate cartilage and in endochondral ossification. Proc Natl Acad Sci U S A 101:17192–17197
10. Stickens D, Behonick DJ, Ortega N, Heyer B, Hartenstein B, Yu Y, Fosang AJ, Schorpp-Kistner M, Angel P, Werb Z (2004) Altered endochondral bone development in matrix metalloproteinase 13-deficient mice. Development 131:5883–5895
11. Zelzer E, Glotzer DJ, Hartmann C, Thomas D, Fukai N, Soker S, Olsen BR (2001) Tissue specific regulation of VEGF expression during bone development requires Cbfa1/Runx2. Mech Dev 106:97–106
12. Mackie EJ, Ahmed YA, Tatarczuch L, Chen KS, Mirams M (2008) Endochondral ossification: how cartilage is converted into bone in the developing skeleton. Int J Biochem Cell Biol 40:46–62
13. Song F, Li B, Stocum DL (2010) Amphibians as research models for regenerative medicine. Organogenesis 6:141–150
14. Glowacki J (1998) Angiogenesis in fracture repair. Clin Orthop Relat Res 355:S82–S89
15. Young HE, Bailey CF, Dalley BK (1983) Gross morphological analysis of limb regeneration in postmetamorphic adult *Ambystoma*. Anat Rec 206:295–306
16. Einhorn TA, Gerstenfeld LC (2015) Fracture healing: mechanisms and interventions. Nat Rev Rheumatol 11:45–54

17. Bais M, McLean J, Sebastiani P, Young M, Wigner N, Smith T, Kotton DN, Einhorn TA, Gerstenfeld LC (2009) Transcriptional analysis of fracture healing and the induction of embryonic stem cell-related genes. PLoS One 4, e5393
18. Mitogawa K, Makanae A, Satoh A, Satoh A (2015) Comparative analysis of cartilage marker gene expression patterns during axolotl and *Xenopus* limb regeneration. PLoS One 10, e0133375
19. Kragl M, Knapp D, Nacu E, Khattak S, Maden M, Epperlein HH, Tanaka EM (2009) Cells keep a memory of their tissue origin during axolotl limb regeneration. Nature 460:60–69
20. Echeverri K, Clarke JDW, Tanaka EM (2001) In vivo imaging indicates muscle fiber dedifferentiation is a major contributor to the regenerating tail blastema. Dev Biol 236:151–164
21. Echeverri K, Tanaka EM (2002) Ectoderm to mesoderm lineage switching during axolotl tail regeneration. Science 298:1993–1996
22. Schnapp E, Kragl M, Rubin L, Tanaka EM (2005) Hedgehog signaling controls dorsoventral patterning, blastema cell proliferation and cartilage induction during axolotl tail regeneration. Development 132:3243–3253
23. Cosden RS, Lattermann C, Romine S, Gao J, Voss SR, MacLeod JN (2011) Intrinsic repair of full-thickness articular cartilage defects in the axolotl salamander. Osteoarthr Cart 19:200–205
24. Lozito TP, Tuan RS (2015) Lizard tail regeneration: regulation of two distinct cartilage regions by Indian hedgehog. Dev Biol 399:249–262
25. Fisher RE, Geiger LA, Stroik LK, Hutchins ED, George RM, Denardo DF, Kusumi K, Rawls JA, Wilson-Rawls J (2012) A histological comparison of the original and regenerated tail in the green anole, *Anolis carolinensis*. Anat Rec 295:1609–1619
26. Hutchins ED, Markov GJ, Eckalbar WL, George RM, King JM, Tokuyama MA, Geiger LA, Emmert N, Ammar MJ, Allen AN, Siniard AL, Corneveaux JJ, Fisher RE, Wade J, DeNardo DF, Rawls JA, Huentelman MJ, Wilson-Rawls J, Kusumi K (2014) Transcriptomic analysis of tail regeneration in the lizard *Anolis carolinensis* reveals activation of conserved vertebrate developmental and repair mechanisms. PLoS One 9, e105004
27. Alibardi L (2015) Regeneration of articular cartilage in lizard knee from resident stem/progenitor cells. Int J Mol Sci 16:20731–20747
28. Seifert AW, Kiama SG, Seifert MG, Goheen JR, Palmer TM, Maden M (2012) Skin shedding and tissue regeneration in African spiny mice (Acomys). Nature 489:561–565
29. Fitzgerald J, Rich C, Burkhardt D, Allen J, Herzka AS, Little CB (2008) Evidence for articular cartilage regeneration in MRL/MpJ mice. Osteoarthr Cart 16:1319–1326
30. Clark LD, Clark RK, Heber-Katz E (1998) A new murine model for mammalian wound repair and regeneration. Clin Immunol Immunopathol 88:35–45
31. Lozito TP, Tuan RS (2016) Lizard tail regeneration as an instructive model of enhanced healing capabilities in an adult amniote. Connect Tissue Res 1–10
32. Angus d'A Bellairs SVB, Darevsky IS, Hubert J, Kupriyanova LA, Pieau C, Raynaud A, Saint Girons H, Shine R, Uzzell T, Yaron Z (1985) The biology of the Reptilia. Wiley, New York
33. Shargal E, Rath-Wolfson L, Kronfeld N, Dayan T (1999) Ecological and histological aspects of tail loss in spiny mice (Rodentia: Muridae, *Acomys*) with a review of its occurrence in rodents. J Zool 249:187–193
34. Dubost G, Gasc JP (1987) The process of total tail autotomy in the South-American rodent, *Proechimys*. J Zool 212:563–572
35. Heydemann A (2012) The super super-healing MRL mouse strain. Front Biol 7:522–538
36. Gourevitch DL, Clark L, Bedelbaeva K, Leferovich J, Heber-Katz E (2009) Dynamic changes after murine digit amputation: the MRL mouse digit shows waves of tissue remodeling, growth, and apoptosis. Wound Repair Regen 17:447–455
37. Bedelbaeva K, Snyder A, Gourevitch D, Clark L, Zhang XM, Leferovich J, Cheverud JM, Lieberman P, Heber-Katz E (2010) Lack of p21 expression links cell cycle control and appendage regeneration in mice. Proc Natl Acad Sci U S A 107:5845–5850
38. Johnson DH, Pedowitz RA (2007) Practical orthopaedic sports medicine and arthroscopy [electronic resource]. Wolters Kluwer Health/Lippincott Williams & Wilkins, Philadelphia

39. Rhee DK, Marcelino J, Baker M, Gong Y, Smits P, Lefebvre V, Jay GD, Stewart M, Wang H, Warman ML, Carpten JD (2005) The secreted glycoprotein lubricin protects cartilage surfaces and inhibits synovial cell overgrowth. J Clin Invest 115:622–631
40. Waller KA, Zhang LX, Elsaid KA, Fleming BC, Warman ML, Jay GD (2013) Role of lubricin and boundary lubrication in the prevention of chondrocyte apoptosis. Proc Natl Acad Sci U S A 110:5852–5857
41. Pritchard JJ, Ruzicka AJ (1950) Comparison of fracture repair in the frog, lizard and rat. J Anat 84:236–261
42. Callaghan JJ (2003) The adult knee, vol 1. Lippincott Williams & Wilkins, Philadelphia, PA
43. Brooks P (2003) Inflammation as an important feature of osteoarthritis. Bull World Health Organ 81:689–690
44. World Health Organization (2002) The world health report 2002: reducing risks, promoting healthy life. World Health Organization, Geneva
45. Goldring MB (2006) Update on the biology of the chondrocyte and new approaches to treating cartilage diseases. Best Pract Res Clin Rheumatol 20:1003–1025
46. Lethbridge-Cejku M, Helmick CG, Popovic JR (2003) Hospitalizations for arthritis and other rheumatic conditions: data from the 1997 National Hospital Discharge Survey. Med Care 41:1367–1373
47. Felson DT (2004) An update on the pathogenesis and epidemiology of osteoarthritis. Radiol Clin North Am 42:1–9
48. Knutsen G, Drogset JO, Engebretsen L, Grontvedt T, Isaksen V, Ludvigsen TC, Roberts S, Solheim E, Strand T, Johansen O (2007) A randomized trial comparing autologous chondrocyte implantation with microfracture. Findings at five years. J Bone Joint Surg Am 89:2105–2112
49. Saris DB, Vanlauwe J, Victor J, Haspl M, Bohnsack M, Fortems Y, Vandekerckhove B, Almqvist KF, Claes T, Handelberg F, Lagae K, van der Bauwhede J, Vandenneucker H, Yang KG, Jelic M, Verdonk R, Veulemans N, Bellemans J, Luyten FP (2008) Characterized chondrocyte implantation results in better structural repair when treating symptomatic cartilage defects of the knee in a randomized controlled trial versus microfracture. Am J Sports Med 36:235–246
50. Bentley G, Biant LC, Vijayan S, Macmull S, Skinner JA, Carrington RW (2012) Minimum ten-year results of a prospective randomised study of autologous chondrocyte implantation versus mosaicplasty for symptomatic articular cartilage lesions of the knee. J Bone Joint Surg (Br) 94:504–509
51. Mason C, Dunnill P (2008) A brief definition of regenerative medicine. Regen Med 3:1–5
52. Kuo CK, Li WJ, Mauck RL, Tuan RS (2006) Cartilage tissue engineering: its potential and uses. Curr Opin Rheumatol 18:64–73
53. O'Driscoll SW, Keeley FW, Salter RB (1986) The chondrogenic potential of free autogenous periosteal grafts for biological resurfacing of major full-thickness defects in joint surfaces under the influence of continuous passive motion. An experimental investigation in the rabbit. J Bone Joint Surg Am 68:1017–1035
54. Grande DA, Singh IJ, Pugh J (1987) Healing of experimentally produced lesions in articular cartilage following chondrocyte transplantation. Anat Rec 218:142–148
55. Grande DA, Pitman MI, Peterson L, Menche D, Klein M (1989) The repair of experimentally produced defects in rabbit articular cartilage by autologous chondrocyte transplantation. J Orthop Res 7:208–218
56. Brittberg M, Lindahl A, Nilsson A, Ohlsson C, Isaksson O, Peterson L (1994) Treatment of deep cartilage defects in the knee with autologous chondrocyte transplantation. N Engl J Med 331:889–895
57. Jiang YZ, Zhang SF, Qi YY, Wang LL, Ouyang HW (2011) Cell transplantation for articular cartilage defects: principles of past, present, and future practice. Cell Transplant 20:593–607
58. Peterson L, Vasiliadis HS, Brittberg M, Lindahl A (2010) Autologous chondrocyte implantation: a long-term follow-up. Am J Sports Med 38:1117–1124

59. Cole BJ, DeBerardino T, Brewster R, Farr J, Levine DW, Nissen C, Roaf P, Zaslav K (2012) Outcomes of autologous chondrocyte implantation in study of the treatment of articular repair (STAR) patients with osteochondritis dissecans. Am J Sports Med 40:2015–2022
60. Zaslav K, Cole B, Brewster R, DeBerardino T, Farr J, Fowler P, Nissen C (2009) A prospective study of autologous chondrocyte implantation in patients with failed prior treatment for articular cartilage defect of the knee: results of the Study of the Treatment of Articular Repair (STAR) clinical trial. Am J Sports Med 37:42–55
61. Bartlett W, Skinner JA, Gooding CR, Carrington RW, Flanagan AM, Briggs TW, Bentley G (2005) Autologous chondrocyte implantation versus matrix-induced autologous chondrocyte implantation for osteochondral defects of the knee: a prospective, randomised study. J Bone Joint Surg (Br) 87:640–645
62. Ahmed TA, Hincke MT (2010) Strategies for articular cartilage lesion repair and functional restoration. Tissue Eng Part B Rev 16:305–329
63. Filardo G, Kon E, Roffi A, Di Martino A, Marcacci M (2013) Scaffold-based repair for cartilage healing: a systematic review and technical note. Arthroscopy 29:174–186
64. Vanlauwe JJ, Claes T, Van Assche D, Bellemans J, Luyten FP (2012) Characterized chondrocyte implantation in the patellofemoral joint: an up to 4-year follow-up of a prospective cohort of 38 patients. Am J Sports Med 40:1799–1807
65. Tuan RS, Boland G, Tuli R (2003) Adult mesenchymal stem cells and cell-based tissue engineering. Arthritis Res Ther 5:32–45
66. Teo BJ, Buhary K, Tai BC, Hui JH (2013) Cell-based therapy improves function in adolescents and young adults with patellar osteochondritis dissecans. Clin Orthop Relat Res 471: 1152–1158
67. Kasemkijwattana C, Hongeng S, Kesprayura S, Rungsinaporn V, Chaipinyo K, Chansiri K (2011) Autologous bone marrow mesenchymal stem cells implantation for cartilage defects: two cases report. J Med Assoc Thail 94:395–400
68. Wakitani S, Mitsuoka T, Nakamura N, Toritsuka Y, Nakamura Y, Horibe S (2004) Autologous bone marrow stromal cell transplantation for repair of full-thickness articular cartilage defects in human patellae: two case reports. Cell Transplant 13:595–600
69. Nejadnik H, Hui JH, Feng Choong EP, Tai BC, Lee EH (2010) Autologous bone marrow-derived mesenchymal stem cells versus autologous chondrocyte implantation: an observational cohort study. Am J Sports Med 38:1110–1116
70. Li QA, Tang JC, Wang RY, Bei CY, Xin LW, Zeng YJ, Tang XY (2011) Comparing the chondrogenic potential in vivo of autogeneic mesenchymal stem cells derived from different tissues. Artif Cells Blood Substit Immobil Biotechnol 39:31–38
71. Koga H, Muneta T, Nagase T, Nimura A, Ju YJ, Mochizuki T, Sekiya I (2008) Comparison of mesenchymal tissues-derived stem cells for in vivo chondrogenesis: suitable conditions for cell therapy of cartilage defects in rabbit. Cell Tissue Res 333:207–215
72. Murphy JM, Fink DJ, Hunziker EB, Barry FP (2003) Stem cell therapy in a caprine model of osteoarthritis. Arthritis Rheum 48:3464–3474
73. Kayakabe M, Tsutsumi S, Watanabe H, Kato Y, Takagishi K (2006) Transplantation of autologous rabbit BM-derived mesenchymal stromal cells embedded in hyaluronic acid gel sponge into osteochondral defects of the knee. Cytotherapy 8:343–353
74. Guo X, Park H, Young S, Kretlow JD, van den Beucken JJ, Baggett LS, Tabata Y, Kasper FK, Mikos AG, Jansen JA (2010) Repair of osteochondral defects with biodegradable hydrogel composites encapsulating marrow mesenchymal stem cells in a rabbit model. Acta Biomater 6:39–47
75. Yan H, Yu CL (2007) Repair of full-thickness cartilage defects with cells of different origin in a rabbit model. Arthroscopy 23:178–187
76. Im GI, Kim DY, Shin JH, Hyun CW, Cho WH (2001) Repair of cartilage defect in the rabbit with cultured mesenchymal stem cells from bone marrow. J Bone Joint Surg (Br) 83:289–294
77. Park JS, Yang HN, Woo DG, Chung HM, Park KH (2009) In vitro and in vivo chondrogenesis of rabbit bone marrow-derived stromal cells in fibrin matrix mixed with growth factor loaded in nanoparticles. Tissue Eng Part A 15:2163–2175

78. Zhou G, Liu W, Cui L, Wang X, Liu T, Cao Y (2006) Repair of porcine articular osteochondral defects in non-weightbearing areas with autologous bone marrow stromal cells. Tissue Eng 12:3209–3221
79. Sharma B, Williams CG, Khan M, Manson P, Elisseeff JH (2007) In vivo chondrogenesis of mesenchymal stem cells in a photopolymerized hydrogel. Plast Reconstr Surg 119:112–120
80. Park J, Gelse K, Frank S, von der Mark K, Aigner T, Schneider H (2006) Transgene-activated mesenchymal cells for articular cartilage repair: a comparison of primary bone marrow-, perichondrium/periosteum- and fat-derived cells. J Gene Med 8:112–125
81. Yoshimura H, Muneta T, Nimura A, Yokoyama A, Koga H, Sekiya I (2007) Comparison of rat mesenchymal stem cells derived from bone marrow, synovium, periosteum, adipose tissue, and muscle. Cell Tissue Res 327:449–462
82. Xie X, Wang Y, Zhao C, Guo S, Liu S, Jia W, Tuan RS, Zhang C (2012) Comparative evaluation of MSCs from bone marrow and adipose tissue seeded in PRP-derived scaffold for cartilage regeneration. Biomaterials 33:7008–7018
83. Fernandez FB, Shenoy S, Babu SS, Varma HK, John A (2012) Short-term studies using ceramic scaffolds in lapine model for osteochondral defect amelioration. Biomed Mater 7, 035005
84. Lee JC, Min HJ, Park HJ, Lee S, Seong SC, Lee MC (2013) Synovial membrane-derived mesenchymal stem cells supported by platelet-rich plasma can repair osteochondral defects in a rabbit model. Arthroscopy 29:1034–1046
85. Lee JC, Lee SY, Min HJ, Han SA, Jang J, Lee S, Seong SC, Lee MC (2012) Synovium-derived mesenchymal stem cells encapsulated in a novel injectable gel can repair osteochondral defects in a rabbit model. Tissue Eng Part A 18:2173–2186
86. Suzuki S, Muneta T, Tsuji K, Ichinose S, Makino H, Umezawa A, Sekiya I (2012) Properties and usefulness of aggregates of synovial mesenchymal stem cells as a source for cartilage regeneration. Arthritis Res Ther 14:R136
87. Pei M, He F, Boyce BM, Kish VL (2009) Repair of full-thickness femoral condyle cartilage defects using allogeneic synovial cell-engineered tissue constructs. Osteoarthritis Cartilage 17:714–722
88. Pei M, He F, Li J, Tidwell JE, Jones AC, McDonough EB (2013) Repair of large animal partial-thickness cartilage defects through intraarticular injection of matrix-rejuvenated synovium-derived stem cells. Tissue Eng Part A 19:1144–1154
89. Mifune Y, Matsumoto T, Takayama K, Ota S, Li H, Meszaros LB, Usas A, Nagamune K, Gharaibeh B, Fu FH, Huard J (2013) The effect of platelet-rich plasma on the regenerative therapy of muscle derived stem cells for articular cartilage repair. Osteoarthritis Cartilage 21:175–185
90. Matsumoto T, Cooper GM, Gharaibeh B, Meszaros LB, Li G, Usas A, Fu FH, Huard J (2009) Cartilage repair in a rat model of osteoarthritis through intraarticular transplantation of muscle-derived stem cells expressing bone morphogenetic protein 4 and soluble Flt-1. Arthritis Rheum 60:1390–1405
91. Wakitani S, Aoki H, Harada Y, Sonobe M, Morita Y, Mu Y, Tomita N, Nakamura Y, Takeda S, Watanabe TK, Tanigami A (2004) Embryonic stem cells form articular cartilage, not teratomas, in osteochondral defects of rat joints. Cell Transplant 13:331–336
92. Toh WS, Lee EH, Guo XM, Chan JK, Yeow CH, Choo AB, Cao T (2010) Cartilage repair using hyaluronan hydrogel-encapsulated human embryonic stem cell-derived chondrogenic cells. Biomaterials 31:6968–6980
93. Dattena M, Pilichi S, Rocca S, Mara L, Casu S, Masala G, Manunta L, Manunta A, Passino ES, Pool RR, Cappai P (2009) Sheep embryonic stem-like cells transplanted in full-thickness cartilage defects. J Tissue Eng Regen Med 3:175–187
94. Hiramatsu K, Sasagawa S, Outani H, Nakagawa K, Yoshikawa H, Tsumaki N (2011) Generation of hyaline cartilaginous tissue from mouse adult dermal fibroblast culture by defined factors. J Clin Invest 121:640–657
95. Vinardell T, Sheehy EJ, Buckley CT, Kelly DJ (2012) A comparison of the functionality and in vivo phenotypic stability of cartilaginous tissues engineered from different stem cell sources. Tissue Eng Part A 18:1161–1170

96. Pendleton C, Li Q, Chesler DA, Yuan K, Guerrero-Cazares H, Quinones-Hinojosa A (2013) Mesenchymal stem cells derived from adipose tissue vs bone marrow: in vitro comparison of their tropism towards gliomas. PLoS One 8, e58198
97. Hildner F, Albrecht C, Gabriel C, Redl H, van Griensven M (2011) State of the art and future perspectives of articular cartilage regeneration: a focus on adipose-derived stem cells and platelet-derived products. J Tissue Eng Regen Med 5:E36–E51
98. Yamashita A, Nishikawa S, Rancourt DE (2010) Identification of five developmental processes during chondrogenic differentiation of embryonic stem cells. PLoS One 5, e10998
99. Yamashita A, Liu S, Woltjen K, Thomas B, Meng G, Hotta A, Takahashi K, Ellis J, Yamanaka S, Rancourt DE (2013) Cartilage tissue engineering identifies abnormal human induced pluripotent stem cells. Sci Rep 3:1978
100. Goldring MB, Tsuchimochi K, Ijiri K (2006) The control of chondrogenesis. J Cell Biochem 97:33–44
101. Tuan RS (2003) Cellular signaling in developmental chondrogenesis: N-cadherin, Wnts, and BMP-2. J Bone Joint Surg Am 85-A(Suppl 2):137–141
102. Shum L, Coleman CM, Hatakeyama Y, Tuan RS (2003) Morphogenesis and dysmorphogenesis of the appendicular skeleton. Birth Defects Res C Embryo Today 69:102–122
103. Wezeman FH (1998) Morphological foundations of precartilage development in mesenchyme. Microsc Res Tech 43:91–101
104. Akiyama H, Lefebvre V (2011) Unraveling the transcriptional regulatory machinery in chondrogenesis. J Bone Miner Metab 29:390–395
105. Hwang NS, Varghese S, Lee HJ, Zhang Z, Ye Z, Bae J, Cheng L, Elisseeff J (2008) In vivo commitment and functional tissue regeneration using human embryonic stem cell-derived mesenchymal cells. Proc Natl Acad Sci U S A 105:20641–20646
106. Barry F, Boynton RE, Liu B, Murphy JM (2001) Chondrogenic differentiation of mesenchymal stem cells from bone marrow: differentiation-dependent gene expression of matrix components. Exp Cell Res 268:189–200
107. Byers BA, Mauck RL, Chiang IE, Tuan RS (2008) Transient exposure to transforming growth factor beta 3 under serum-free conditions enhances the biomechanical and biochemical maturation of tissue-engineered cartilage. Tissue Eng Part A 14:1821–1834
108. Huang AH, Stein A, Tuan RS, Mauck RL (2009) Transient exposure to transforming growth factor beta 3 improves the mechanical properties of mesenchymal stem cell-laden cartilage constructs in a density-dependent manner. Tissue Eng Part A 15:3461–3472
109. Estes BT, Wu AW, Guilak F (2006) Potent induction of chondrocytic differentiation of human adipose-derived adult stem cells by bone morphogenetic protein 6. Arthritis Rheum 54:1222–1232
110. Hennig T, Lorenz H, Thiel A, Goetzke K, Dickhut A, Geiger F, Richter W (2007) Reduced chondrogenic potential of adipose tissue derived stromal cells correlates with an altered TGFbeta receptor and BMP profile and is overcome by BMP-6. J Cell Physiol 211:682–691

Chapter 5
Digit Regeneration in Mammals

Lindsay A. Dawson, Paula P. Schanes, Connor Dolan, Paulina Ketcham, Ling Yu, Mingquan Yan, Tao Li, and Ken Muneoka

5.1 Introduction

The astonishing regenerative ability of the urodele amphibian limb has long been investigated as the chief model for regeneration in vertebrates. The urodele limb responds to amputation via a process called epimorphic regeneration, and involves the formation of a blastema comprised of proliferating cells that are undifferentiated. The regeneration process involves a series of stages (e.g. inflammation, wound closure, dedifferentiation, cell migration, etc.) many of which are known to be essential for the successful replacement of the amputated structure. Such a stepwise view of regeneration [1] points to the fact that the regenerative response involves a complex series of interconnecting processes, and not simply an event that can be toggled on or off. While the urodele limb represents a beacon for regeneration among higher vertebrates, mammals, including humans, are not without regenerative capabilities and can successfully regenerate the distal portion of the fingertip [2, 3]. This regeneration response is amputation level specific, in that conservative treatment of amputations distal to the nail matrix can successfully regenerate, while amputations proximal to the nail matrix results in a more traditional wound healing response that culminates with scar formation. This regenerative ability is particularly enhanced in

L.A. Dawson (✉) • P.P. Schanes • C. Dolan • P. Ketcham • L. Yu • M. Yan • K. Muneoka (✉)
Department of Veterinary Physiology & Pharmacology, Texas A&M University, College Station, TX 77843-4466, USA
e-mail: ldawson@cvm.tamu.edu; pschanes@cvm.tamu.edu; cpdolan@cvm.tamu.edu; pketcham1990@gmail.com; lyu@cvm.tamu.edu; myan@cvm.tamu.edu; kmuneoka@cvm.tamu.edu

T. Li
Department of Hand Surgery, Wuhan Union Hospital, Tongji Medical College, Huazhong University of Science and Technology, 1277 Jiefang Avenue, Wuhan, Hubei, P.R. China, 430022
e-mail: lili97801@sohu.com

J. Wilson-Rawls, K. Kusumi (eds.), *Innovations in Molecular Mechanisms and Tissue Engineering*, Stem Cell Biology and Regenerative Medicine,
DOI 10.1007/978-3-319-44996-8_5

children, however a similar response has been documented in adults [4]. While human fingertip regeneration is well documented in the clinical literature, the details of this response have not been well characterized, thus it remains more a curiosity rather than a model upon which regenerative therapies might evolve.

Like the human fingertip, the digit tip of mice possesses a similar ability to regenerate; amputation through the terminal phalanx results in the faithful restoration of the digit tip [5]. This regenerative response is highly reproducible and occurs following digit amputation during fetal, neonatal and adult stages. Like the regenerating urodele limb, the digit tip regenerative response involves blastema formation and goes through a series of inter-dependent stages, some of which are known to be essential for successful regeneration. Regeneration of the mouse digit tip correlates with amputation distal to the nail matrix making it amputation level specific, much like human fingertip regeneration. Thus, the regenerating mouse mammalian digit functions as an important investigative model for uncovering the details surrounding endogenous mammalian regenerative response, and conversely, regenerative failure. In this chapter we summarize recent advances in our understanding of how regeneration in mammals is controlled.

5.2 Overview of Digit Regeneration

The adult mouse distal phalanx (P3) is a triangular shaped bone with a relatively wide base and a comparatively smaller distal apex, encased dorsally and medially within the nail organ (Fig. 5.1a, b). Distal amputation of P3 transects multiple tissue types, including the P3 bone, the surrounding soft connective tissue, nerves, vasculature, the ventral epidermis, the nail and associated nail bed (Fig. 5.1b, c). Distal amputation does not remove the nail matrix, damage the ventrally located fat pad, or transect the highly vascularized marrow cavity (Fig. 5.1c). While prompt wound closure is associated with the urodele regeneration response, the wound closure of the mammalian P3 digit is comparatively slow and shows considerable variability, taking between 8 and 12 days to complete [6]. Wound healing is complicated because the wound epidermis does not close directly over the amputated bone surface. Instead, the epidermis initially retracts and attaches to the periosteal surface at a location proximal to the original amputation plane (Fig. 5.1d). During this initial wound response and prior to wound closure, overt tissue histolysis occurs, exemplified by the degradation of the bone stump (Fig. 5.1d). The bone degradation response is mediated by osteoclasts that create large pits in the stump bone that eventually causes a secondary amputation proximal to the original (Fig. 5.1e, f). Following this injury-induced secondary amputation, the proximally contracted epidermis is able to migrate through the region of regressed bone and forms a wound epidermis that eventually caps the distal digit region. In most cases the degradation of distal bone is incomplete and the completion of wound closure results in

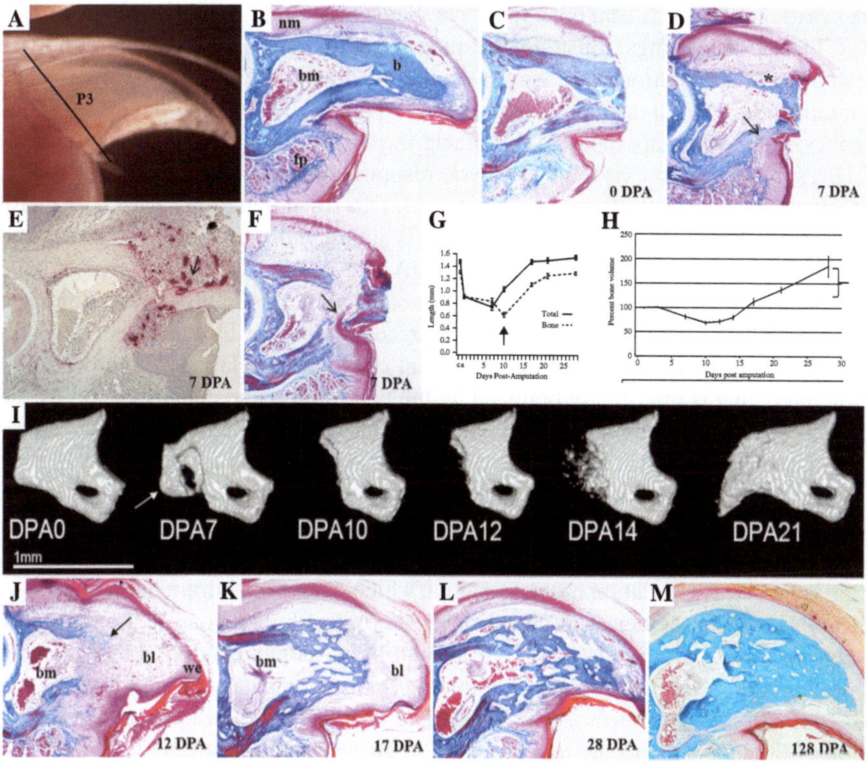

Fig. 5.1 (**a**) Photograph of the unamputated mouse digit illustrating the nail encasing the triangular shaped P3 bone. (**b**) Histological section of the unamputated digit, showing the nail matrix (nm), the P3 bone (**b**), the bone marrow (bm), and the fat pad (fp). (**c**) Amputation removes the distal digit tip without transecting the vasculature rich marrow cavity. (**d**) Simultaneous events at 7 DPA include epidermal retraction from the amputation plane (*arrow*) and osteoclast erosion (*asterisk*) of the bone stump proximal to the original amputation level. (**e**) TRAP staining at 7 DPA detects osteoclasts adjacent to the eroded bone stump (*arrow*) and secondary amputation plane. (**f**) Epidermal migration (*arrow*) beneath the eroded bone stump at 7 DPA. (**g–i**) Quantification of bone length, volume, and μCT 3-D renderings illustrate the regeneration response is characterized by an initial degradation of bone (*arrows* in **g** and **i**) and a corresponding decrease in bone length and volume, followed by continued bone regeneration resulting in an overall overshoot in bone volume. (**j**) The variable process of wound closure and wound epidermis (WE) formation occurs between 8 and 12 DPA. At 12 DPA, the bone marrow cavity is open to the wound site and the blastema (bl) is prominent, and regeneration of bone is evident in the proximal blastema (*arrow*). (**k**) By 17 DPA, proximal regeneration of woven bone (**b**) has enclosed the bone marrow cavity, and the distal blastema is still present. (**l**) AT 28 DPA, the amputated digit is largely regenerated, showing evidence of robust woven bone formation, soft connective tissue regeneration, and reestablishment of pre-amputation bone and digit length. (**m**) At 128 DPA, the regenerated digit is comprised of thick trabecular bone, and consequently is not a perfect replica of the unamputated digit. (**a–f** and **j–m**) Distal is to the *right*, dorsal is to the *top*. (**i**) Distal is to the *left*, dorsal is to the *top*. (**a–g** and **j–m**) are reprinted from Fernando et al. [6] and (**h**) and (**i**) are reprinted from Sammarco et al. [7]

the casting off of the distal bone fragment. Sequential micro-computed tomography (μCT) 3-D renderings illustrate the intense degradation of the bone stump in response to amputation, and showcases the eventual expelling of the distal bone stump evident by 10 days post-amputation (DPA; Fig. 5.1i). In summary, the early stages of this amputation model is characterized by a slow and variable wound healing response and an extensive histolytic response of bone tissue that results in an injury-induced re-amputation of the digit [6]. This re- amputation causes a decrease in both bone volume and bone length that is significantly greater than that caused by the initial amputation injury (Fig. 5.1g, h).

The timing of wound closure and blastema formation is tightly linked; the blastema forms rapidly once wound closure is complete. The blastema consists of a population of undifferentiated mesenchymal cells with a relatively high proliferation index that is proximally bounded by the bone stump and distally bound by the thickened wound epidermis (Fig. 5.1j). One consequence of the injury-induced re-amputation is that the distal wound site where blastema formation occurs is directly adjacent to the P3 marrow region which is highly vascularized. We note that the transition between the cell dense blastema and the highly vascularized but relatively less cell dense bone marrow is quite dramatic (Fig. 5.1j). The next step of regeneration is the redifferentiation of the blastema, which occurs via intramembranous ossification, with no evidence of chondrogenesis [6, 8, 9]. The bone redifferentiation stage begins at approximately 12 DPA, with initial boney condensations showing continuity with the bone stump at both the proximal boundary of the blastema and the dorsal periosteal surface (Fig. 5.1j). The overt intramembranous redifferentiation of the blastema occurs in a proximal to distal fashion, resulting in an increase in bone and associated surrounding connective tissue length, as well as an increase in bone volume (Fig. 5.1h, k, l). By 28 DPA, the digit has completed regeneration, including integration of the newly formed bone with the bone stump, reconstitution of the marrow cavity, and regeneration of the surrounding connective tissues, i.e. vasculature, dermis, epidermis, and nail. The digit regenerates to the pre-amputation bone length and characteristic pointed edge, but notably, the resulting regenerate exhibits a disorganized trabecular bone pattern and an overshoot in bone volume (Fig. 5.1l). By 128 DPA, the trabecular bone of the regenerate has condensed, yet the relative disorganized morphology is easily distinguishable from the original bone stump, thus the regeneration response does not result in a perfect replica of the amputated structure (Fig. 5.1m).

While digit tip regeneration in adult mice digit embodies a rather complex but coordinated series of events leading to blastema formation and re-differentiation, the events associated with neonatal and fetal regeneration appear to be less complex. Neonatal digits are structurally patterned but immature, containing cells still undergoing chondrogenesis and ossification is just initiating [8]. Unlike adult digit amputations, the neonatal wound epidermis closes directly over the stump bone, however the timing to completely close the wound is highly variable by comparison to non-regenerative amputations [10]. The osteoclast-mediated bone degradation response observed in adult amputations is absent in neonates, and regenerative outgrowth is not preceded by an injury-induced re-amputation of the stump. Once formed the neonatal blastema is similar to the adult blastema, composed of undifferentiated mesenchymal cells with a relatively high proliferative index and reduced vascularity. Re-differentiation of the digit tip occurs in a proximal to distal manner by direct ossification as it does in adults. Regeneration of the fetal digit tip is, however, considerably different. The fetal mouse digit tip is largely undifferentiated at the time of amputation, wound closure and blastema formation occurs rapidly, and the regeneration process is completed over a very short timeframe. While uniquely different from post-natal and adult regeneration, the process of fetal regeneration can occur ex vivo making it an attractive model to experimentally dissect the regenerative response [11].

5.3 Regenerative Failure and Induced Regeneration

Mouse digit tip regeneration is a valuable model for mammalian regeneration, but it also has added value because the response is amputation level specific; amputation at any digit or limb level proximal to the digit tip fails to regenerate. Uncovering critical events important for the endogenous regenerative response can be studied during the healing events associated with regenerative failure. In recent years strategies to induce regeneration from proximal amputations have been successful, thus demonstrating for the first time, that mammalian regeneration can be specifically induced and, in some cases, provide evidence that events associated with the non-regenerating healing response actively inhibit the regenerative process. These findings are critically important because they demonstrate the existence of a regenerative potential that is actively repressed in normally non-regenerative injuries that can be activated by targeted treatments during wound healing.

One of the first successful demonstrations of induced mammalian regeneration focused on the importance of Bone Morphogenetic Proteins (BMPs), members of the TGFβ superfamily of signaling molecules, signaling in digit tip regeneration [12]. Studies of the regenerating fetal digit demonstrated that expression of the homeobox-containing gene *Msx1* and *Bmp4* were co-expressed in the digit tip and were re-expressed during the regenerative response [11, 13]. Moreover, the expression domain of *Msx1* correlates with regeneration permissive amputation levels, and amputation studies of *Msx1* mutant digits failed to regenerate, thus providing evidence that the *Msx1* gene is functionally required for a regenerative response [11]. In other studies, *Msx1* was shown to repress cell differentiation during embryonic development [14], and was implicated in the control of tail regeneration in the Xenopus tadpole [15]. In addition, *Msx1* activity has been linked to the induction of de-differentiation of mammalian myotubes in vitro [16], suggesting that its activity is essential for a regenerative response. Using the digit regeneration defect, Han et al. [11] discovered that treatment with exogenous BMP4 was able to rescue the *Msx1* regeneration phenotype in a dose dependent manner. This resulted in enhanced cell proliferation, the re-expression of a number of digit specific genes, and the restoration of the digit tip. To test if BMP signaling was critical for wildtype digit tip regeneration, treatment of amputated wildtype digits with Noggin, a BMP signaling antagonist, was found to inhibit the regeneration response [11]. These regeneration studies identified both *Msx1* function and BMP signaling as key players in the control of mammalian digit regeneration. Just as important, analogous studies focused on two other genes expressed in the fetal digit tip, *Msx2* and *Dlx5*, failed to yield a regeneration phenotype as single mutants or in combination indicating that regenerative failure is not simply a general phenotype of distally expressed digit genes [11, 17].

Shifting to neonatal digit studies, *Msx1* and *Bmp4* are also prominently expressed in the neonatal digit tip at the time of amputation, and re-expressed during digit tip regeneration [8, 18]. In this post-natal regeneration model, digit tip regeneration is inhibited by targeting Noggin treatment to the amputation wound after epidermal closure indicating that BMP signaling is essential for the regenerative response [10]. This finding indicates that BMP signaling is critical for a successful regenerative response. In addition to *Bmp4*, the neonate digit regeneration blastema is associated with prominent expression of *Bmp2* and *Bmp7* as well as known receptors for BMP signaling (*Bmpr1a*, *Bmpr1b*, and *Bmpr2*) so the specific BMP signal(s) important for regeneration remains uncertain. To determine whether BMP signaling could induce a regenerative response, studies were focused on amputations transecting the proximal portion of the neonate P3 bone. Such amputations do not form a blastema, but undergo a wound healing response and resulted in truncated digits [10]. To test for the induction of a regenerative response, purified BMP2, BMP4, or BMP7 were absorbed onto agarose micro-carrier beads and implanted between the wound epidermis and the digit stump after wound closure was complete. These tests showed that BMP2 and BMP7, but not BMP4, were able to induce a regenerative response that resulted in the restoration of the amputated digit tip. The BMP-induced regeneration response was associated with the accumulation of a blastema of proliferating undifferentiated cells that expressed two blastema-specific marker genes, *Msx1* and *Pedf* (see below). Notably, whereas the endogenous regeneration of P3 forms a blastema that re-differentiates bone by intramembranous ossification [8], BMP-induced regeneration created a blastema that formed an endochondral ossification center contiguous with the bone stump [10]. Moreover, the polarity of the induced endochondral ossification center was the same as the proximal P3 growth plate with proliferating chondrocytes proximal to the distal hypertrophic chondrocytes (Fig. 5.2a, b). Taken together, proximal P3 amputation injuries respond to BMP treatment via the formation of a transient blastema, comprised of proliferating cells re-expressing relevant genes, reactivating a position-specific differentiation response that results in the regeneration of a normally patterned terminal phalanx (Fig. 5.2e).

A similar study was conducted to test the regeneration potential after amputation midway through the neonate middle phalanx (P2). The P2 bone is similar to a typical long bone with joint articulations both proximally (P1/P2 joint) and distally (P2/P3 joint), and a marrow region that extends the length of the bone. After amputation, wound closure occurs over the amputated bone and is consistently completed within 4–5 days [19]. BMP2 soaked micro-carrier beads were implanted between the wound epidermis and the amputated bone stump after wound closure. The anatomical

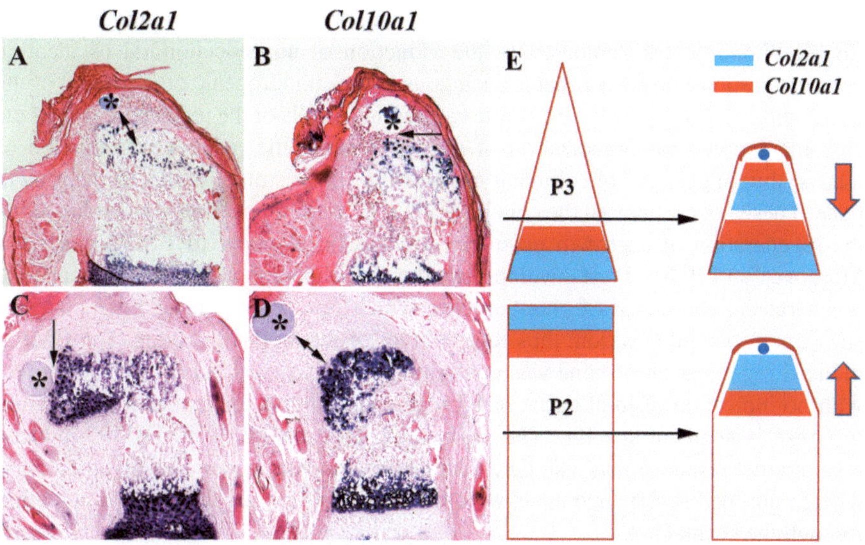

Fig. 5.2 (**a**) In-situ hybridization probing for the cartilage proliferation marker *Col2a1* in proximally amputated P3 digits show transcripts within the regenerating tissue, yet not in close association (*arrowheads*) with the BMP2 soaked bead (*asterisk*). (**b**) In situ hybridization probing for the hypertrophic cartilage marker *Col10a1* in proximally amputated P3 digits show transcripts in direct association (*arrow*) with the BMP2 soaked bead (*asterisk*). (**c**) In situ hybridization for *Col2a1* in P2 amputated digits treated with BMP2 shows robust transcript localization directly adjacent (*arrow*) to the bead (*asterisk*). (**d**) *Col10a1* in situ hybridization detected transcripts in the regenerating P2 stump, yet not in close association (*arrowheads*) with the BMP2 soaked bead (*asterisk*). (**e**) Schematic diagram illustrating the in-situ results following BMP2-induced regeneration of proximal P3 amputation and P2 amputation. Amputation level shown as *arrow*. The BMP2-induced P3 regeneration response is characterized by blastema formation and the subsequent differentiation into chondrocytes to regenerate the digit in a distal to proximal fashion, thus following P3 developmental mechanisms. The BMP2-induced P2 regeneration response is characterized by blastema formation and the regeneration of a growth plate, resulting in proximal to distal bone regeneration, therefore following the mechanism of P2 development. Distal is to the *top*. Reprinted from Yu et al. [19]

response to BMP2 was the elongation of the P2 skeletal element to almost 90 % of stage-matched unamputated control length, however there is no evidence for the regeneration of the P2/P3 joint or any part of the P3 element. Thus, the induced response was specific to the amputated skeletal segment. The BMP2-induced regeneration response included the re-expression of blastema marker genes (*Msx1* and *Pedf*), an enhanced proliferation response, and the accumulation of mesenchymal cells associated with the bead and distal to the amputation plane. This mesenchymal cell aggregate initially expressed *Col2a1* indicative of chondrocyte differentiation, and then transitioned to *Col10a1* expression indicative of differentiation into hypertrophic chondrocytes. The organization of this induced endochondral ossification center was consistent with an elongating skeletal element: proliferating chondrocytes distal to the hypertrophic chondrocytes associated with the stump bone, and resulted in the deposition of new bone onto the amputated stump (Fig. 5.2c, d). The spatial organization of distally-located proliferating chondrocytes differentiating into hypertrophic chondrocytes demonstrates the induction of an endochondral ossification center in response to BMP2 that is analogous to the chondrogenic organization of the developing P2 bone (Fig. 5.2e). While cartilaginous cells of the regenerate lacked the customary columnar organization of a growth plate, the polarized production of matrix, thus suggestive of columnar chondrocytes, was similar to that of the growth plate. These data supported the conclusion that P2 bone elongation was mediated via the regeneration of a growth plate-like structure in response to BMP2 treatment. Moreover, and akin to bone development, the endochondral ossification center of the regenerate is comprised of chondrocytes, which function to create a template for subsequent bone deposition, thus regeneration of the amputated bone. The BMP2-induced regeneration of bone was contiguous with and constrained to P2, however with no indication of joint tissue or P3 regeneration, thus the induced regeneration response is segment specific. The segment-specific nature of this BMP2-induced regenerative response was also demonstrated in an adult model of limb amputation where patterned skeletal elements were induced to regenerate following amputation through the shank [19].

For non-regenerating amputations of adult digits, the P2 amputation has emerged as a model system to study both tissue repair after amputation and induced regeneration [20–24]. P2 is analogous to other long bones of the mammalian body, in that it cannot mount a successful endogenous regeneration response after amputation. Instead, the injured structure undergoes wound repair and ultimately scar formation, nonetheless, the wound repair response of P2 is quite dynamic. P2 is located centrally within the digit, bounded dorsally by a ligament and overlying dermis and epidermis rich with hair follicles, and ventrally by a deep digital flexor tendon with associated fibrocartilage, and underlying dermis and epidermis. The P3 marrow is distinct from the P2 marrow, in that P3 contains abundant vasculature and is relatively more cell dense compared to the fatty P2 marrow. Amputation through P2 exposes the marrow cavity to the wound site and completely removes the nail organ and digit tip (Fig. 5.3a). P2 wound closure is achieved by forward contraction of the dermal tissues, apparent by 6 DPA, distinct from the characteristic epidermal retraction of P3 post amputation (Fig. 5.3b). Moreover, while osteoclast mediated bone erosion is associated with the P2 amputation response, evident by the significant dip in bone volume at 7 DPA, the bone erosion typically does not result in expelling the bone with concomitant wound closure as it does in P3. Instead the epidermis and underlying soft connective tissue migrate distal to the bone stump, closing the wound, with wound epidermis formation by 9 DPA (Fig. 5.3c). Unlike the blastemal intramembranous bone redifferentiation that represents P3 regeneration, the P2 response to amputation is via the formation of a periosteal-derived cartilaginous callus, testing immunopositive for several cartilage matrix proteins, including Collagen 2 and Aggrecan by 9 DPA (Fig. 5.3c–e outlined). Importantly, the transient cartilaginous callus is observed exclusively along the periosteal surface proximal to

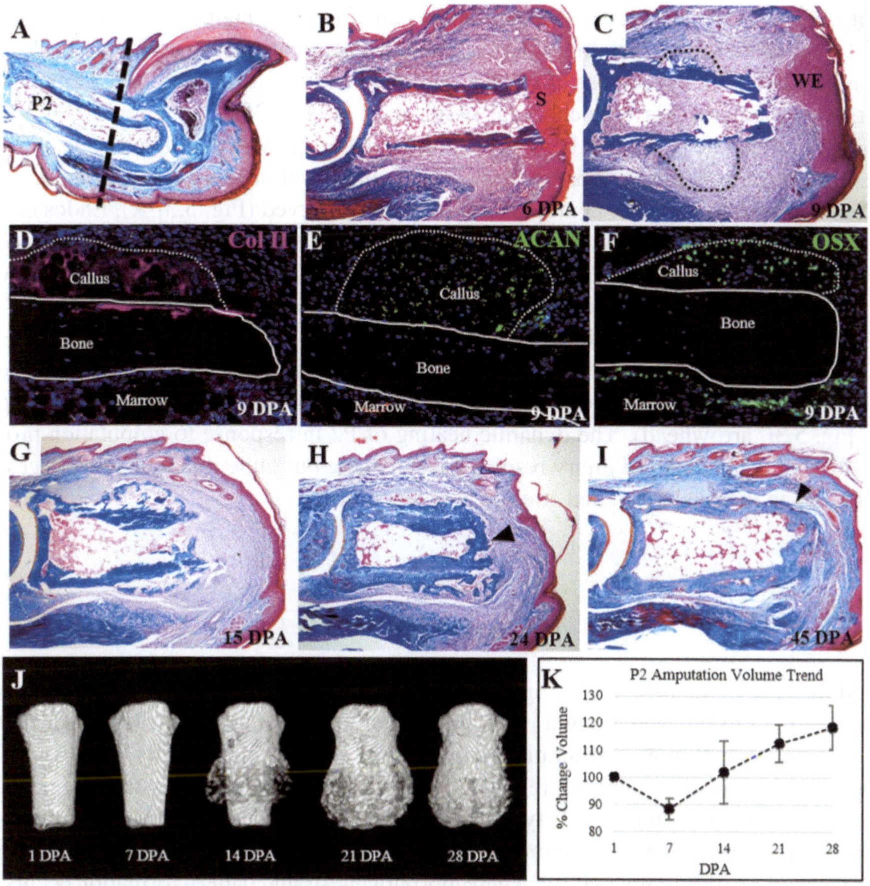

Fig. 5.3 (**a**) P2 amputation (*dashed line*) transects the mid-portion of the bone and completely removes the nail organ. (**b**) Distal scab formation (S) and circumferential swelling of the digit are apparent by 6 DPA. (**c**) By 9 DPA, wound closure over the distal stump and wound epidermis formation (WE) have occurred. Robust chondrogenesis along the periosteal surface (*outlined*) extends perpendicular to the bone stump, not distal to the amputation plane. (**d**) ColII immunostaining localized to the periosteal cartilaginous callus at 9 DPA. (**e**) ACAN immunostaining confirms the presence of a periosteal chondrogenic callus at 9 DPA. (**f**) Immunostaining for the osteoblast marker Osterix, OSX, shows osteoblasts localized to the 9 DPA periosteal callus and within the bone marrow space. (**g**) By 15 DPA, the cartilaginous callus has been largely replaced with woven bone and marrow formation. (**h**) Distal intramembranous bone formation has sealed the marrow space (*arrowhead*) and the callus has undergone remodeling by 24 DPA. (**i**) At 45 DPA, the amputated digit has completely healed and the ventral tendon has re-inserted into the bone on the dorsal surface (*arrowhead*). (**j** and **k**) μCT 3-D renderings and quantification of bone volume changes illustrate the robust periosteal growth in response to amputation, corresponding to an overshoot in bone volume, yet no increase in bone length. (**a–i**) Distal is to the *right*, dorsal is to the *top*. (**j**) Distal is to the *bottom*. Reprinted from Dawson et al. [23]

the bone stump, not distal to the amputation plane. The cartilaginous callus functions as a template for the invading osteoblasts, which act to deposit woven bone in close association with the periosteal surface, thus the bone repair response to P2 amputation is mediated via endochondral ossification (Fig. 5.3f). Sequential μCT 3-D images illustrate periosteal bone deposition by 14 DPA, with continued bone deposition corresponding to an increase in bone volume, yet new bone formation distal to the amputation plane, i.e. regeneration, is not observed (Fig. 5.3j, k). Endosteal-derived osteoblasts function to cap the bone stump, separating the marrow space from the wound site by 24 DPA (Fig. 5.3h, arrowhead). Remodeling of the bone stump continues, complete with reconstitution of the marrow, evident at 24 and 45 DPA (Fig. 5.3h, i). In addition, there is clear evidence that soft tissues, such as the ventral tendon which normally attaches to the P3 element, elongates across the amputation wound and re-inserts into the dorsal aspect of the P2 bone stump (Fig. 5.3i, arrowhead). The dynamic healing of P2 in response to amputation provides evidence that the injury response is not static but quite dynamic and suggestive of an initiated but failed attempt at regeneration.

5.4 The Blastema

At the core of the problem of regenerative failure is the inability to form a blastema rather than to proceed along a pathway of traditional wound repair. It is clear that cells present at non-regenerating amputation wounds individually possess the potential for involvement in a regeneration response, but what is missing is a mechanism whereby different cell types can interact to organize a coordinated multi-tissue response. In regeneration-competent models the blastema functions as the regenerative developmental site where morphogenesis and pattern formation occurs. In addition, there is experimental evidence that patterning mechanisms guiding development are similar to those guiding regeneration [25, 26]. For this reason, understanding how a regeneration blastema forms following amputation can provide critical clues for inducing or otherwise enhancing regenerative capabilities. The first question about blastema formation concerns the origin of cells. The blastema is a heterogeneous population of undifferentiated cells that are proliferative and have the potential to differentiate into the structures that make up the digit tip; tissue types include bone, bone marrow, connective tissue, blood vessels, nerves and epidermis. The question of cell origin can be addressed using cell type specific markers coupled with a lineage marker. There are a number of important questions that need to be addressed and some progress has been made in the past few years. These questions include what tissue types contribute cells? Do cells arise from stem/progenitors and/or from dedifferentiation? Are cells lineage restricted or can they transdifferentiate during the re-differentiation process? Recent studies using genetic approaches to label specific cell types and track them during digit regeneration provide evidence for lineage-restriction in regeneration [18, 27, 28]. These studies show that (1) epidermal cells gave rise to the nail organ and wound epidermis, (2) endothelial cells contribute only to the vasculature of the regenerate, (3) osteoblasts precursors of the limb give rise only to regenerated bone, and (4) as a negative result, transplanted hematopoietic stem cells do not contribute to the regenerate. These finding are consistent with the conclusion that cells involved in the regeneration of the urodele limb are lineage-restricted [29], so it is tempting to draw the general conclusion that the regeneration process does not involve the reprogramming of cell types [18, 27]. Yet, it is important to point out that a number of cell types have yet to be carefully tested and there is considerable evidence that support the participation of multipotent progenitor cells in other injury repair models [30]. The question of whether progenitor cells arise from a population of stem cells versus the de-differentiation of mature cells has yet to be addressed in this mammalian regenerative response.

The second critical question to address involves how cells are recruited to form the blastema at the amputation wound site. During regeneration, cells migrate from different regions of the amputation wound to form the centrally located blastema [31, 32]. Previous studies have implicated Stromal Cell Derived Factor-1 (SDF-1)/CXCR4 signaling in the cellular recruitment to sites of BMP2-induced ectopic bone formation, bone repair, and zebrafish fin regeneration [33–35]. A comprehensive study investigating the role of SDF-1/CXCR4 signaling in blastema formation and digit regeneration was carried out [36] and a summary of this study follows. Immunohistochemical and in situ hybridization studies focused on the regenerating digit identified cells expressing SDF-1 in the blastema, wound epidermis, and bone marrow vasculature, and other cells expressing CXCR4 in the wound epidermis and within the blastema (Fig. 5.4a–d). Immunostaining for Phospho-CXCR4, used to identify the SDF-1 mediated activation of CXCR4, showed positive signal localized to the wound epidermis and the vasculature proximal to the blastema (Fig. 5.4e). In line with this, CD31+ endothelial cells within the blastema tested immunopositive for SDF-1, suggesting endothelial cells may function in cellular recruitment during regeneration (Fig. 5.4f). Primary cultures of blastema cells express CXCR4 and display a dose-dependent response to SDF1 in transwell migration assays that is inhibited by AMD3100, a known antagonist for CXCR4 signaling. To test the in vivo role of SDF-1/CXCR4 signaling in neonatal digit regeneration, systemic injections of AMD3100 resulted in a significant attenuation in regenerated bone length compared to vehicle control treated digits (Fig. 5.4g). Importantly, no difference was found in bone length of the developing P3 digit after AMD3100 systemic treatment, providing evidence that SDF-1/CXCR4 signaling is specific to the regeneration response. The BMP2-induced regeneration response was used to provide a gain of function test for a role of SDF1 recruitment in regeneration. First, introducing labeled blastema cells in conjunction with a BMP2 bead resulted in a chemotactic response with labeled cells aggregating around the BMP2 source. Transwell migration assays demonstrated that BMP2 itself does not influence blastema cell migration indicating that the in vivo response was indirect. BMP2 treatment of P2 amputations was found to induce both SDF-1 and CXCR4 expression and SDF-1 expression co-localized with endothelial cells in the amputation wound (Fig. 5.4h–m). A BMP2/SDF-1 link was confirmed in vitro by studies showing that BMP2

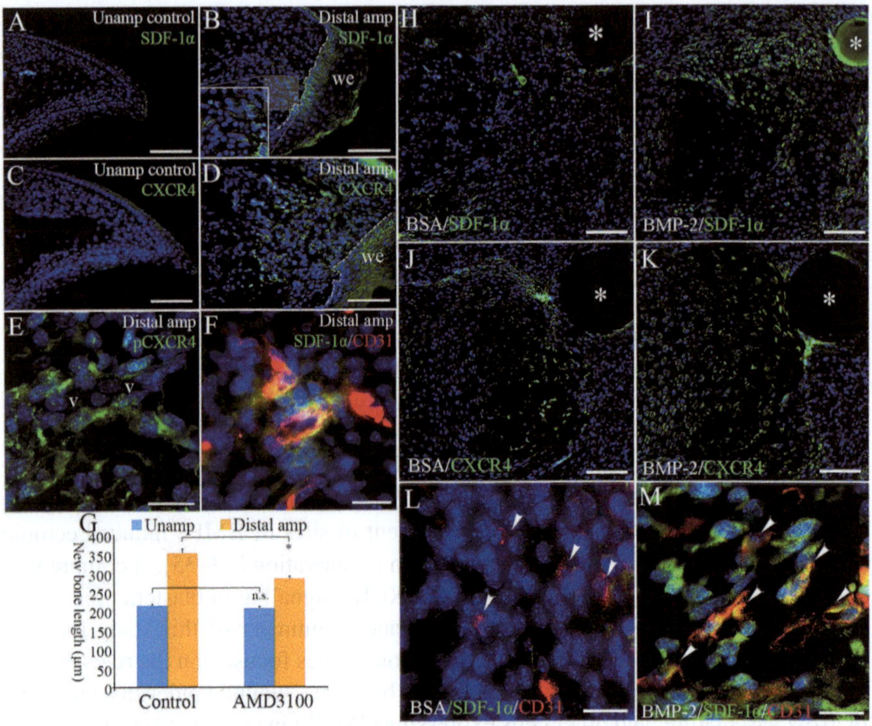

Fig. 5.4 (**a–f** and **h–m**) Immunostained digits, counterstained with DAPI. (**a**) SDF-1α is not detected in the unamputated digit. (**b**) P3 amputation induces SDF1α in the proximal wound epidermis (WE) and within the blastema (inset), shown at 7 DPA. (**c**) CXCR4 is not detected in the unamputated digit. (**d**) CXCR4 is localized to the WE and the blastema of the regenerating 7 DPA P3 digit. (**e**) Activated CXCR4, detected by phospho-CXCR4 immunostaining, is present in the newly formed vessels (V) of the proximal blastema at 7 DPA. (**f**) Co-immunostaining for SDF-1α and the endothelial cell marker CD31 show double labeled cells within the regenerating digit at 7 DPA. (**g**) Inhibition of the SDF1α/CXCR4 signaling axis via AMD3100 systemic treatment significantly attenuates the bone regeneration response compared to vehicle treated digits. (**h** and **j**) P2 amputation and BSA control bead (*asterisk*) treatment show a low level of SDF1α and CXCR4 protein at 2 DPI. (**i** and **k**) BMP2-treated P2 amputations show heightened SDF1α and CXCR4 immunostaining in close association with the bead at 2 DPI. (**l** and **m**) Co-immunostaining for SDF1α and CD31+ endothelial cells (*arrowheads*) after P2 amputation and bead treatment shows double labeled cells in BMP2-treated samples at 2 DPI. Distal is to the *right*, dorsal is to the *top*. Reprinted from Lee et al. [17, 36]

treatment of human microvascular endothelial cells (HMVEC) stimulated a dose-dependent increase in *Sdf-1* transcripts, and that BMP2-treated HMVEC cells stimulated migration of blastema cells that was specifically inhibited by AMD3100. Finally, engraftment of COS cells over-expressing *Sdf1* into the neonatal P2 amputation induced a partial skeletal elongation response that is similar to the BMP2 induced response. These studies provide both loss and gain of function evidence that cell recruitment via the SDF-1/CXCR4 signaling axis plays a critical role in blastema cell recruitment during regeneration. In addition, endothelial cells in the blastema and at non-regenerating amputation wounds are identified as critical mediators of BMP2 action by transducing the BMP2 signal into a functional cell recruitment signal.

The third critical topic concerns one of the hallmarks of the regenerative process–the proliferative nature of the blastema. Endogenous blastema formation as well as BMP-induced regeneration is associated with enhanced proliferation. Using a transgenic BMP reporter mouse strain, Yu et al. [19] showed that BMP2 induced P2 regeneration specifically enhanced proliferation of BMP responsive cells suggesting that the mitogenic action of BMP2 was a rapid and direct effect on cells at the amputation wound. In addition to BMP2, there is evidence that regenerative proliferation is also controlled by WNT signaling that is linked to a neurotrophic effect critical for the regeneration response [28].

The amputation level-specific nature of digit tip regeneration is associated with the presence of the nail matrix in the stump. The nail organ is a continuously elongating structure that consists of a nail plate that encases a proximal nail matrix of proliferating nail stem cells, a distal nail matrix of transiently amplifying cells, and the nail bed that extends to the distal digit tip [28]. Nail elongation is a process that requires canonical WNT signaling, and nail dysmorphogenesis results when this signaling pathway is disrupted [28, 37]. Importantly, regenerative defects are also observed when the canonical WNT signaling pathway is disrupted, thus providing an explanation for the close link between regenerative capabilities and the nail in both humans and mice [28, 37]. In addition, studies in which the canonical WNT pathway is constitutively activated in epidermal cells, including nail cells, show that regeneration following amputation at a proximal P3 level can be induced [28], thus canonical WNT signaling by epidermal cells is required for the endogenous regeneration response, and can induce regeneration from a normally non-regenerative amputation injury. A secondary feature of canonical WNT signaling in epidermal cells is evidence that inhibiting signaling causes a reduction in innervation associated with the regeneration response [28]. This raises the possibility that the nail matrix effect on digit regeneration may be mediated via modification of a neurotrophic effect on the regenerative response. The effect of denervation on cell proliferation and blastema formation in regenerating salamander limbs is well documented, resulting in the complete inhibition of regeneration [38]. Unlike the salamander limb, however, denervation of the mouse digit tip does not completely inhibit the regenerative response, but it does impair the normal formation of both nail and bone of the regenerate [28, 39, 40]. Denervation of the digit tip inhibits mesenchymal proliferation and the expression of *Fgf2* which is mitogenic for blastema cells both in vivo and in vitro [28], thus providing evidence that the FGF signaling pathway is also an important trophic influence for mammalian regeneration. While there is likely a laundry list of trophic influences important for the digit regenerative response, the evidence to date point to two critical signaling pathways (BMP and WNT) that are required for the endogenous regenerative response, and can also induce regeneration at a non- regenerative amputation.

5.5 Wound to Blastema

One of the prominent characteristics of the blastema is its relative avascularity compared to surrounding tissues. This is true for both the salamander limb blastema as well as the mouse digit blastema [6, 41]. The digit blastema is not devoid of endothelial cells, instead they are present as individual cells dispersed randomly throughout the blastema and not organized into functional vascular units. The endothelial cells of the blastema express the stem cell marker Sca1, suggesting that the regeneration of vasculature in the digit tip is mediated via endothelial stem cells [6]. Associated with avascularity, the wound healing and early blastema stages is characterized by the expression of the anti- angiogenic factor, *pigment epithelial derived factor* (*Pedf*) (Fig. 5.5a) and the absence of transcripts for the angiogenic factor, vascular endothelial growth factor A (*VegfA*) [42, 43]. The expression of the antiangiogenic factor *Pedf* and the corresponding lack of expression of the angiogenic factor *Vegf*, coupled with the presence of endothelial stem cells in the blastema, are consistent with the conclusion that blastemal avascularity is causally linked to the control of angiogenesis during blastema formation. Why is blastema avascularity important? To test the importance of re-vascularization in regeneration, amputated neonate digits were treated with VEGF-soaked beads after the completion of wound closure. VEGF-treatment led to an increase in endothelial cells within the blastema and adjacent to the VEGF source within 3 days post implantation (DPI), an enhanced level of vascularity by 7 DPI and the complete inhibition of the regenerative response (Fig. 5.5b, c) [42]. The inhibition of regeneration by VEGF was shown to be dose-dependent, and these findings provide evidence that the control of re-vascularization during blastema formation plays a critical role in the control of regeneration.

The conclusion that enhanced re-vascularization inhibited regeneration is suggestive that the avascular state of the blastema is required for the regenerative response. To address this issue we explored the role that PEDF played in the inhibition of re-vascularization. We took advantage of the observation that treatment of the digit amputation with BMP9, a member of the TGFβ superfamily, inhibited the regenerative response (Fig. 5.5h). While *Bmp9* is not expressed during digit tip regeneration, BMP9 has been shown to act in a context-dependent manner acting as either an anti-angiogenic factor or to enhance angiogenesis [44–46], and a microarray analysis of BMP9 treated digits showed a significant increase in several known modulators of angiogenesis, including *Vegfa*. Further studies showed that BMP9 treatment induced an immediate upregulation of *Vegfa* expression, and a persistent and expanded *Vegfa* expression domain at later time points (Fig. 5.5d–g), and an enhanced level of re-vascularization. In short, the data were consistent with the conclusion that the BMP9 inhibition of regeneration was causally linked to the induced expression of *Vegfa*, and we were able to use this model to test the role of PEDF in blastema formation and regeneration. PEDF treatment after BMP9 inhibition markedly decreased the expression domain of *Vegfa*, thus restoring the avascular character of the blastema, and ultimately rescuing osteogenesis associated with the regeneration response (Fig. 5.5h–n). Taken together, the evidence shows PEDF

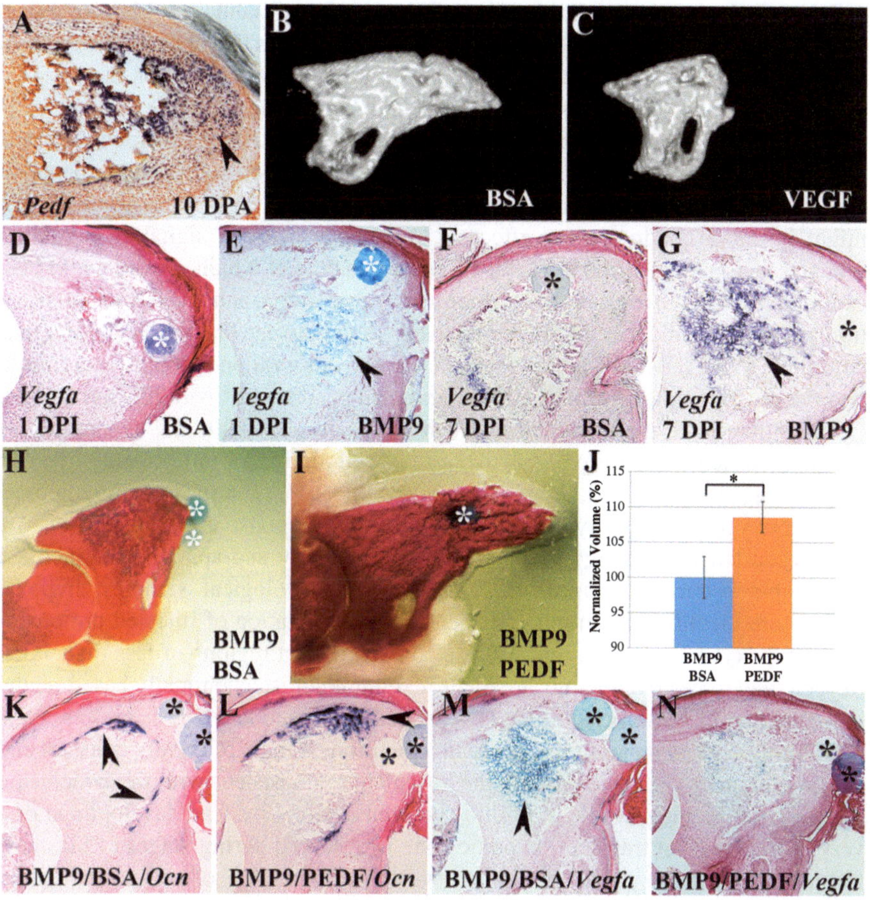

Fig. 5.5 (**a**) Transcripts for *Pedf* are found within the stump and the blastema (*arrowhead*) of the regenerating 10 DPA digit. (**b** and **c**) μCT 3-D renderings at 14 DPI illustrate the inhibition of regeneration in VEGF-treated digits. (**d** and **e**) *Vegfa* transcripts are absent in BSA treated control digits, yet are detected (*arrowhead*) at 1 DPI in close association with the BMP9 soaked bead (*asterisk*). (**f** and **g**) By 7 DPI, *Vegfa* transcripts are detected in the proximal region of the BSA treated digit, yet show an expanded expression domain (*arrowhead*) in BMP9 treated digits. (**h**–**j**) Application of PEDF in conjunction with BMP9 treatment rescues the BMP9-induced inhibition of regeneration versus BMP9/BSA treated digits, resulting in a significant increase in bone regeneration. (**k** and **l**) At 7 DPI, BMP9/BSA treated digits show attenuated *Ocn* expression, while BMP9/PEDF-treated digits show an expanded domain of *Ocn* expression (*arrowheads*). (**m** and **n**) At 7 DPI, BMP9/BSA treated digits show an expanded domain of *Vegfa* expression, while BMP9/PEDF treated digits show a paucity of *Vegfa* transcripts. Distal is to the *right*, dorsal is to the *top*. Reprinted from Yu et al. [42]

functions during the early stages of regeneration to repress VEGF induced revascularization and to maintain the blastema in an avascular state. The data are consistent with the conclusion that an avascular blastema is essential for promoting the regenerative response.

In general, tissue vascularity plays a critical role in maintaining physiological function and homeostasis. Although there are a few tissues that are inherently avascular, e.g. cornea and articular cartilage, most of the body depends on the vasculature for distributing nutrients and blood bound signals as well as removing waste products. In this context, the avascular digit blastema exists as a structure that is largely isolated from the physiological influences of the body. One consequence of this is that the blastema creates a physically less turbulent microenvironment that might be more conducive for effective long range cell-cell signaling involving secreted factors (e.g. BMPs, WNTs, FGFs, etc.), some of which are known to play essential roles in the regenerative response. In this context, enhanced revascularization might physically disrupt intercellular signaling between blastema cells and thus contribute to the failed regenerative response. The observation that blastema formation occurs following VEGF or BMP9 treatment is consistent with this hypothesis. On the other hand, the avascular microenvironment also limits the availability of essential nutrients, such as oxygen, to blastema cells and this would create a hypoxic microenvironment. Indeed, recent studies document that the blastema is hypoxic, and that oxygen availability during the regenerative response is dynamic [7].

Oxygen tensions change dynamically in temporally and spatially distinct and predictable patterns during P3 regeneration. In histological samples, hypoxic regions are identified by immunohistochemical localization of injected pimonidazole (Hypoxyprobe-1 Plus) that forms stable adducts in regions of less than 1.3 % oxygen, and hyperoxic regions are identified immunohistochemically based on the presence of FBLX5, a protein that is stabilized at oxygen levels greater than 5.5 % [7]. During digit regeneration, hyperoxic conditions remain relatively constant and are predominantly associated with the vasculature, consistent with the conclusion that vasculature plays a role in limiting oxygen availability. The development of a very prominent, but transient, hypoxic zone is observed during stages of blastema formation, and that zone dissipates with the initiation of re-differentiation (Fig. 5.6a–d). To test the requirement of the hypoxic blastema microenvironment on the regeneration process, mice were exposed to Hyperbaric Oxygen (HBO) treatment, targeting the period of blastema formation. A single HBO treatment is sufficient to disrupt the hypoxic microenvironment of the blastema, but regeneration is not inhibited by either targeted or continuous HBO treatment [7, 47], thus the hypoxic microenvironment of the blastema is not required for successful regeneration. However, there is clear indication that HBO treatment does induce specific modifications of the regeneration process. HBO treatment enhances the activity of osteoclasts during the histolytic phase resulting in an extended period of bone degradation and a delay in blastema formation (Fig. 5.6e) [47]. This suggests that while a hypoxic blastema is not a requirement for regeneration, cells involved in regeneration are responsive to changing oxygen tension and this plays a role in regulating phase transitions during the regenerative process. The interaction between osteoclasts and osteoblasts has been studied in the context of bone turnover and bone diseases, such as osteoporosis and osteopetrosis, and regulatory pathways have been identified. Osteoclasts are derived from monocytes and express Receptor Activator of Nuclear Factor kβ (RANK), while osteoclastogenesis during

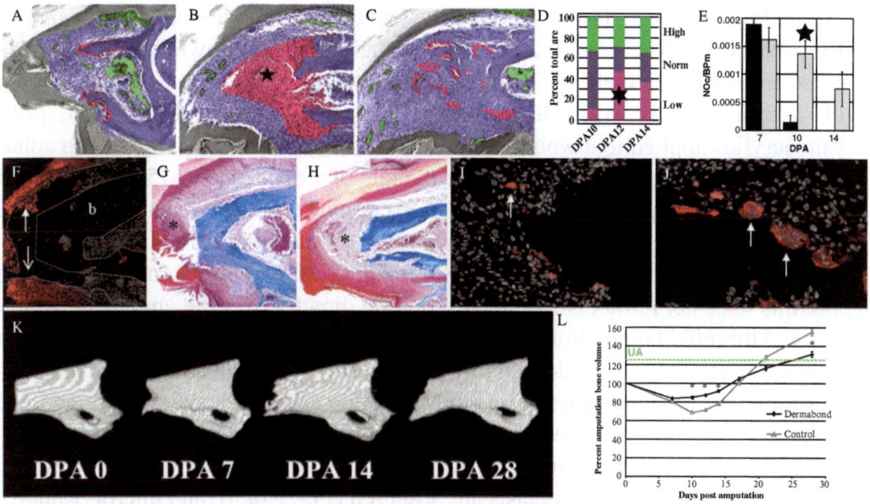

Fig. 5.6 P3 regeneration represents a dynamic oxygen microenvironment. (**a–d**) Oxygen pseudo-shading of in vivo Hypoxyprobe-1 Plus staining (<1.3 % oxygen) indicates hypoxic microenvironments (<1.3 % oxygen) at (**a**) 10 DPA (*shading* encircling the bone stumps distally), (**b**) 12 DPA (*shading* in blastema; *star* emphasizing expanded area), and 14 DPA (isolated regions adjacent to newly forming bone). (**d**) Quantification of Hypoxybrobe-1 as a percentage of total area. (**e**) Effect of daily HBO application on osteoclast numbers at 7, 10, and 14 DPA. *Gray bars* indicate HBO treated digits; *black bars* are controls. Results are expressed as mean ±SEM. *Star* indicates significance. NOc/BPm: number of osteocalsts/bone perimeter. (**f**) Cyanoacrylic wound dressing (Dermabond) application results in a hypoxic epidermis. *Arrow* and *arrowhead* indicate Hypoxyprobe-1 staining in both the dorsal and ventral wound epidermis, respectively. (**g**) Dermabond-treated digits exhibit early wound closure distal to the amputation plane (*asterisk*). (**h**) A representative Dermabond-treated digit illustrating wound closure and blastema formation (*asterisk*) by 6 DPA. (**i** and **j**) Immunostaining for the osteoclast marker Cathepsin K (CathK) shows decreased osteoclast fusion (*arrow*) in Dermabond-treated digits and large multinuclear immunopositive osteoclasts (*arrows*) in untreated digits. Nuclei counterstained with DAPI. (**k**) μCT 3-D renderings in Dermabond-treated digits show attenuated bone degradation in response to Dermabond treatment. Distal is to the *left*, dorsal is to the *top*. (**l**) Quantification of bone volume changes from μCT data of Dermabond treated and control digits, *asterisks* for significance. (**a–d**) are reprinted from Sammarco et al. [7], (**e**) is reprinted from Sammarco et al. [47], and (**f–l**) are reprinted from Simkin et al. [48]

inflammation is induced by RANK Ligand (RANKL) stimulation [49]. Both RANK and RANKL are cell surface receptors and this signaling pathway is down- regulated by Osteoprotegerin (OPG), a secreted decoy receptor for RANK [50]. Regulation of osteoclastogenesis by this pathway is regulated by oxygen sensing mechanism in osteoblasts by responding to hypoxia by enhancing Hypoxia Inducing Factor 2a activity which directly up-regulates expression of OPG [51]. In this way hypoxia induces the termination of an osteoclastogenic response. This model provides a general mechanism in which a blastema hypoxic event can trigger osteoblasts within the blastema to produce OPG thereby inhibiting osteoclast activity and signaling the transition from a degradative phase to an anabolic phase of digit regeneration.

In addition to osteoclasts, cells of the wound epidermis are also responsive to HBO treatments. The completion of epidermal closure following amputation is slowed by HBO treatment [47], and this process can be accelerated by treatment with a commercially available cyanoacrylic wound dressing, Dermabond [48]. Cyanoacrylics applied to a wound rapidly polymerize to form a flexible skin adhesive, which have been shown to enhance the rate of wound closure [52]. When used as a wound dressing for digit tip amputations, application of Dermabond creates a sustained hypoxic microenvironment that is restricted to the stump wound epidermis and doubles the rate of wound closure (Fig. 5.6f, g). The Dermabond treated epidermis does not retract but is able to migrate directly over the amputated stump bone, and this effect is ameliorated by HBO treatment. These data implicate hypoxia as a positive regulator of epidermal migration during wound healing. Once amputation wound closure is complete blastema formation occurs precociously (Fig. 5.6h) and the forming blastema becomes hypoxic. Immunohistochemical studies show that large multinucleated cathepsin K positive osteoclasts are absent, however cathepsin K positive pre-osteoclasts are observed (Fig. 5.6i, j), and microCT analyses show that stump bone degradation is significantly reduced (Fig. 5.6k). The precociously formed blastema re-differentiates to form the distal digit tip, but remarkably the overshoot in regenerated bone that characterizes the endogenous response is not observed, and the regenerated bone has a structure that is lamellar rather than woven. In other words, the resulting regenerate is structurally identical to the amputated digit tip when the wound epidermis is induced to close rapidly. It is interesting to note that limb regeneration in salamanders is characterized by rapid wound closure that is driven by a rapid cell migration response coupled with a general injury induced epidermal swelling response [53, 54].

Data from HBO and Dermabond studies are consistent with a model in which oxygen availability plays a key role in regulating the histolytic phase of regeneration, particularly hypoxia induced termination of osteoclast activity. These studies also demonstrate that regulating osteoclast activity correlates with blastema size: reduced osteoclast activity induced by Dermabond results in small blastemas whereas enhancing osteoclast activity with HBO treatment results in large blastemas. These observations suggest that osteoclast activity is linked to the number or proliferation of stump cells that participate in the regenerative response. One possibility is that histolysis of mature stump tissues is required to release progenitor cells so they can participate in blastema formation; enhancing histolysis results in a larger regeneration competent progenitor cell population while reducing histolysis results in a smaller population of progenitors. An alternative, but not mutually exclusive, possibility is that proteolytic activities associated with histolysis degrades extracellular matrix and releases chemotactic and/or mitogenic signals known to be present in mature tissues. For example, bone tissue is known to store extracellular BMPs [55], which have been shown to be essential for a regenerative response [10, 11]. Similarly, matrix degradation products (e.g. cryptic peptides) and metalloproteinase activity have been implicated in cell recruitment and regeneration after amputation injury [20–22, 56]. Regardless of mechanism, the evidence supports the conclusion that histolysis of mature stump tissue is a critical early phase of a successful regenerative response. Another fascinating aspect of HBO treated regenerates is that when osteoclast mediated bone degradation is de-regulated, the extended phase of bone erosion degrades the stump bone into regions that are normally non-regenerative following simple amputation, yet these digits eventually transition to a blastema phase and regenerate the complete digit tip [47]. In some case we have observed degradation into the P2/P3 joint with the successful regeneration of both joint and distal bone tissues. These observations support the counter-intuitive conclusion that enhancing histolytic degradation of stump tissues can, in fact, enhance regenerative capabilities.

References

1. Endo T, Bryant SV, Gardiner DM (2004) A stepwise model system for limb regeneration. Dev Biol 270:135–145
2. Douglas BS (1972) Conservative management of guillotine amputation of the finger in children. Aust Paediatr J 8:86–89
3. Illingworth CM (1974) Trapped fingers and amputated finger tips in children. J Pediatr Surg 9:853–858
4. Lee LP, Lau PY, Chan CW (1995) A simple and efficient treatment for fingertip injuries. J Hand Surg (Br) 20:63–71
5. Borgens RB (1982) Mice regrow the tips of their foretoes. Science 217:747–750
6. Fernando WA, Leininger E, Simkin J, Li N, Malcom CA, Sathyamoorthi S, Han M, Muneoka K (2011) Wound healing and blastema formation in regenerating digit tips of adult mice. Dev Biol 350:301–310
7. Sammarco MC, Simkin J, Fassler D, Cammack AJ, Wilson A, Van Meter K, Muneoka K (2014) Endogenous bone regeneration is dependent upon a dynamic oxygen event. J Bone Miner Res 29:2336–2345
8. Han M, Yang X, Lee J, Allan CH, Muneoka K (2008) Development and regeneration of the neonatal digit tip in mice. Dev Biol 315:125–135
9. Muller TL, Ngo-Muller V, Reginelli A, Taylor G, Anderson R, Muneoka K (1999) Regeneration in higher vertebrates: limb buds and digit tips. Semin Cell Dev Biol 10:405–413
10. Yu L, Han M, Yan M, Lee EC, Lee J, Muneoka K (2010) BMP signaling induces digit regeneration in neonatal mice. Development 137:551–559
11. Han M, Yang X, Farrington JE, Muneoka K (2003) Digit regeneration is regulated by Msx1 and BMP4 in fetal mice. Development 130:5123–5132
12. Hogan BL (1996) Bone morphogenetic proteins in development. Curr Opin Genet Dev 6:432–438
13. Reginelli AD, Wang YQ, Sassoon D, Muneoka K (1995) Digit tip regeneration correlates with regions of Msx1 (Hox 7) expression in fetal and newborn mice. Development 121:1065–1076
14. Hu G, Lee H, Price SM, Shen MM, Abate-Shen C (2001) Msx homeobox genes inhibit differentiation through upregulation of cyclin D1. Development 128:2373–2384
15. Beck CW, Christen B, Slack JM (2003) Molecular pathways needed for regeneration of spinal cord and muscle in a vertebrate. Dev Cell 5:429–439
16. Odelberg SJ, Kollhoff A, Keating MT (2000) Dedifferentiation of mammalian myotubes induced by msx1. Cell 103:1099–1109
17. Lee J, Corcoran A, Han M, Gardiner DM, Muneoka K (2013) Dlx5 and Msx2 regulate mouse anterior neural tube closure through ephrinA5-EphA7. Dev Growth Differ 55:341–349
18. Lehoczky JA, Robert B, Tabin CJ (2011) Mouse digit tip regeneration is mediated by fate-restricted progenitor cells. Proc Natl Acad Sci U S A 108:20609–20614

19. Yu L, Han M, Yan M, Lee J, Muneoka K (2012) BMP2 induces segment-specific skeletal regeneration from digit and limb amputations by establishing a new endochondral ossification center. Dev Biol 372:263–273
20. Agrawal V, Johnson SA, Reing J, Zhang L, Tottey S, Wang G, Hirschi KK, Braunhut S, Gudas LJ, Badylak SF (2010) Epimorphic regeneration approach to tissue replacement in adult mammals. Proc Natl Acad Sci U S A 107:3351–3355
21. Agrawal V, Kelly J, Tottey S, Daly KA, Johnson SA, Siu BF, Reing J, Badylak SF (2011) An isolated cryptic peptide influences osteogenesis and bone remodeling in an adult mammalian model of digit amputation. Tissue Eng Part A 17:3033–3044
22. Agrawal V, Tottey S, Johnson SA, Freund JM, Siu BF, Badylak SF (2011) Recruitment of progenitor cells by an extracellular matrix cryptic peptide in a mouse model of digit amputation. Tissue Eng Part A 17:2435–2443
23. Dawson LA, Simkin J, Sauque M, Pela M, Palkowski T, Muneoka K (2016) Analogous cellular contribution and healing mechanisms following digit amputation and phalangeal fracture in mice. Regeneration 3:13
24. Turner NJ, Johnson SA, Badylak SF (2010) A histomorphologic study of the normal healing response following digit amputation in C57bl/6 and MRL/MpJ mice. Arch Histol Cytol 73:103–111
25. Muneoka K, Bryant SV (1982) Evidence that patterning mechanisms in developing and regenerating limbs are the same. Nature 298:369–371
26. Muneoka K, Bryant SV (1984) Cellular contribution to supernumerary limbs resulting from the interaction between developing and regenerating tissues in the axolotl. Dev Biol 105:179–187
27. Rinkevich Y, Lindau P, Ueno H, Longaker MT, Weissman IL (2011) Germ-layer and lineage-restricted stem/progenitors regenerate the mouse digit tip. Nature 476:409–413
28. Takeo M, Chou WC, Sun Q, Lee W, Rabbani P, Loomis C, Taketo MM, Ito M (2013) Wnt activation in nail epithelium couples nail growth to digit regeneration. Nature 499:228–232
29. Kragl M, Knapp D, Nacu E, Khattak S, Maden M, Epperlein HH, Tanaka EM (2009) Cells keep a memory of their tissue origin during axolotl limb regeneration. Nature 460:60–65
30. Pang YW, Feng J, Daltoe F, Fatscher R, Gentleman E, Gentleman MM, Sharpe PT (2015) Perivascular stem cells at the tip of mouse incisors regulate tissue regeneration. J Bone Miner Res 31:514–523
31. Gardiner DM, Muneoka K, Bryant SV (1986) The migration of dermal cells during blastema formation in axolotls. Dev Biol 118:488–493
32. Wu Y, Wang K, Karapetyan A, Fernando WA, Simkin J, Han M, Rugg EL, Muneoka K (2013) Connective tissue fibroblast properties are position-dependent during mouse digit tip regeneration. PLoS One 8, e54764
33. Dufourcq P, Vriz S (2006) The chemokine SDF-1 regulates blastema formation during zebrafish fin regeneration. Dev Genes Evol 216:635–639
34. Kitaori T, Ito H, Schwarz EM, Tsutsumi R, Yoshitomi H, Oishi S, Nakano M, Fujii N, Nagasawa T, Nakamura T (2009) Stromal cell-derived factor 1/CXCR4 signaling is critical for the recruitment of mesenchymal stem cells to the fracture site during skeletal repair in a mouse model. Arthritis Rheum 60:813–823
35. Otsuru S, Tamai K, Yamazaki T, Yoshikawa H, Kaneda Y (2007) Bone marrow-derived osteoblast progenitor cells in circulating blood contribute to ectopic bone formation in mice. Biochem Biophys Res Commun 354:453–458
36. Lee J, Marrero L, Yu L, Dawson LA, Muneoka K, Han M (2013) SDF-1alpha/CXCR4 signaling mediates digit tip regeneration promoted by BMP-2. Dev Biol 382:98–109
37. Lehoczky JA, Tabin CJ (2015) Lgr6 marks nail stem cells and is required for digit tip regeneration. Proc Natl Acad Sci U S A 112:13249–13254
38. Pirotte N, Leynen N, Artois T, Smeets K (2016) Do you have the nerves to regenerate? The importance of neural signalling in the regeneration process. Dev Biol 409:4–15
39. Mohammad KS, Neufeld DA (2000) Denervation retards but does not prevent toetip regeneration. Wound Repair Regen 8:277–281

40. Rinkevich Y, Montoro DT, Muhonen E, Walmsley GG, Lo D, Hasegawa M, Januszyk M, Connolly AJ, Weissman IL, Longaker MT (2014) Clonal analysis reveals nerve-dependent and independent roles on mammalian hind limb tissue maintenance and regeneration. Proc Natl Acad Sci U S A 111:9846–9851
41. Mescher AL (1996) The cellular basis of limb regeneration in urodeles. Int J Dev Biol 40:785–795
42. Yu L, Yan M, Simkin J, Ketcham PD, Leininger E, Han M, Muneoka K (2014) Angiogenesis is inhibitory for mammalian digit regeneration. Regeneration 1:33–46
43. Muneoka K, Allan CH, Yang X, Lee J, Han M (2008) Mammalian regeneration and regenerative medicine. Birth Defects Res C Embryo Today 84:265–280
44. David L, Mallet C, Keramidas M, Lamande N, Gasc JM, Dupuis-Girod S, Plauchu H, Feige JJ, Bailly S (2008) Bone morphogenetic protein-9 is a circulating vascular quiescence factor. Circ Res 102:914–922
45. Scharpfenecker M, van Dinther M, Liu Z, van Bezooijen RL, Zhao Q, Pukac L, Lowik CW, ten Dijke P (2007) BMP-9 signals via ALK1 and inhibits bFGF-induced endothelial cell proliferation and VEGF-stimulated angiogenesis. J Cell Sci 120:964–972
46. Suzuki M, Wada H, Yoshino M, Tian L, Shigematsu H, Suzuki H, Alaa M, Tamura H, Fujiwara T, Nagato K, Motohashi S, Moriya Y, Hoshino H, Yoshida S, Shibuya K, Hiroshima K, Nakatani Y, Yoshino I (2010) Molecular characterization of chronic obstructive pulmonary disease-related non-small cell lung cancer through aberrant methylation and alterations of EGFR signaling. Ann Surg Oncol 17:878–888
47. Sammarco MC, Simkin J, Cammack AJ, Fassler D, Gossmann A, Marrero L, Lacey M, Van Meter K, Muneoka K (2015) Hyperbaric oxygen promotes proximal bone regeneration and organized collagen composition during digit regeneration. PLoS One 10, e0140156
48. Simkin J, Sammarco MC, Dawson LA, Tucker C, Taylor LJ, Van Meter K, Muneoka K (2015) Epidermal closure regulates histolysis during mammalian (Mus) digit regeneration. Regeneration 2:14
49. Kitaura H, Kimura K, Ishida M, Kohara H, Yoshimatsu M, Takano-Yamamoto T (2013) Immunological reaction in TNF-alpha-mediated osteoclast formation and bone resorption in vitro and in vivo. Clin Dev Immunol 2013:181849
50. Honma M, Ikebuchi Y, Kariya Y, Suzuki H (2014) Regulatory mechanisms of RANKL presentation to osteoclast precursors. Curr Osteoporos Rep 12:115–120
51. Wu C, Rankin EB, Castellini L, Alcudia JF, LaGory EL, Andersen R, Rhodes SD, Wilson TL, Mohammad KS, Castillo AB, Guise TA, Schipani E, Giaccia AJ (2015) Oxygen-sensing PHDs regulate bone homeostasis through the modulation of osteoprotegerin. Genes Dev 29:817–831
52. Singer AJ, Quinn JV, Hollander JE (2008) The cyanoacrylate topical skin adhesives. Am J Emerg Med 26:490–496
53. Ferris DR, Satoh A, Mandefro B, Cummings GM, Gardiner DM, Rugg EL (2010) Ex vivo generation of a functional and regenerative wound epithelium from axolotl (*Ambystoma mexicanum*) skin. Dev Growth Differ 52:715–724
54. Tanner K, Ferris DR, Lanzano L, Mandefro B, Mantulin WW, Gardiner DM, Rugg EL, Gratton E (2009) Coherent movement of cell layers during wound healing by image correlation spectroscopy. Biophys J 97:2098–2106
55. Urist MR, Huo YK, Brownell AG, Hohl WM, Buyske J, Lietze A, Tempst P, Hunkapiller M, DeLange RJ (1984) Purification of bovine bone morphogenetic protein by hydroxyapatite chromatography. Proc Natl Acad Sci U S A 81:371–375
56. Mu X, Bellayr I, Pan H, Choi Y, Li Y (2013) Regeneration of soft tissues is promoted by MMP1 treatment after digit amputation in mice. PLoS One 8, e59105

[illegible]

Chapter 6
Cellular Approaches to Adult Mammalian Heart Regeneration

Justin Judd and Guo N. Huang

6.1 Introduction

6.1.1 Human Heart Failure

Cardiomyopathies are a major cause of death throughout the world, due in part to the inability of the human heart to significantly regenerate. Improvements in the management of acute myocardial infarction (MI) have led to drastic improvements in short-term mortality rates since the 1960s [1]. However, due to a scarcity of effective long-term therapeutic options, the 5-year survival after diagnosis of heart failure is only 50 % [2]. Thus, heart failure remains an incurable condition and a major cause of death.

The etiology of heart failure is complex, but the syndrome is characterized by cardiac output that is insufficient to meet the metabolic demands of the body. A central complication of heart failure in general is the loss of cardiomyocytes through various cell death mechanisms (reviewed in [3]). In acute myocardial infarction, catastrophic cell death is incurred due to the occlusion of coronary vasculature, which deprives the infarcted region of oxygen and nutrient rich blood. Cardiomyocytes die from both apoptosis and necrosis, though the percent contribution of each death

J. Judd
Cardiovascular Research Institute, University of California San Francisco,
555 Mission Bay Blvd S., SC 382 K, San Francisco, CA 94158, USA
e-mail: Justin.Judd@ucsf.edu

G.N. Huang (✉)
Department of Physiology, Cardiovascular Research Institute, University of California San Francisco, 555 Mission Bay Blvd S., SC 382 K, San Francisco, CA 94158, USA
e-mail: Guo.Huang@ucsf.edu

J. Wilson-Rawls, K. Kusumi (eds.), *Innovations in Molecular Mechanisms and Tissue Engineering*, Stem Cell Biology and Regenerative Medicine,
DOI 10.1007/978-3-319-44996-8_6

mechanism is unclear. Necrotic myocardium is eventually replaced by scar tissue, which lacks the contractile and elastic properties needed for optimal heart function. Ischemic reperfusion is thought to contribute to cell death [4] through inflammation [5], radical oxygen species generation [6], and abnormal calcium handling [7]. Strategies to mitigate peripheral myocardial cell death could potentially be implemented during surgical reperfusion [8, 9]. However, due to the acute lack of blood supply, reperfusion therapy is typically too late to save the dying infarcted myocardium, and fibrotic remodeling follows.

In chronic heart failure, cell death is thought to slowly contribute to deterioration of the ventricular myocardium, thus reducing its ability to effectively contract. This is further complicated in many cases by several aspects of remodeling, such as proliferation of fibroblasts, conversion to myofibroblasts [10], and accompanying alterations in extracellular matrix composition [11]. The re-expression of fetal-specific genes during heart failure has been described by several groups, including a switch from α-myosin heavy chain to β-myosin heavy chain (reviewed in [12]). Metabolic remodeling of cardiomyocytes is also seen in heart failure, such as a shift from fatty acid oxidation to glycolysis (reviewed in [13]). Collectively, these aspects of myocardial remodeling can result in gross morphological changes and associated alterations in tissue mechanics, such as myocardial stiffening, thickening or thinning of the ventricular myocardium, and ventricular dilation, as well as alterations in calcium handling and contractility; all of which can severely affect heart function and feedback on disease progression.

6.1.2 Species Variability in Heart Regeneration

Although adult mammals exhibit an insufficient natural ability to repair damaged myocardium, several lower vertebrates, such as zebrafish, newt, and axolotl, maintain a remarkable regenerative capacity, even in later stages of life. These species-specific differences in regenerative capacity (reviewed in [14, 15]) are an important topic of study in the pursuit of human regeneration. Due to the availability of transgenic models, zebrafish is the best characterized of these species. Mechanistically, genetic lineage tracing experiments show that zebrafish heart regeneration relies primarily on the dedifferentiation and expansion of pre-existing differentiated cardiomyocytes [16, 17]. Poss and colleagues showed this myocardial dedifferentiation involves re-expression of early developmental markers such as *gata4* with an accompanying reduction in myocardial conduction velocity at the injury site [16]. Furthermore, a cryoinjury model demonstrated enhanced cell cycling in a fraction of cardiomyocytes expressing embryonic cardiac myosin heavy chain [18]. Epicardial signaling seems to play a role in the regenerative response to injury [19, 20], but myocyte contributions from epicardial cells directly are apparently limited. The role of a dynamic extracellular matrix was shown to be important in mediating zebrafish heart regeneration [21]. Specifically, fibronectin was upregulated in the myocardium following injury and was required for regeneration. Interestingly, fibronectin deposition in adult mammalian hearts has also been observed post-injury [22, 23], but may signal a fibrotic response in this context [24–26].

Several reports have also demonstrated a strong regenerative ability in adult newt [27–30] and axolotl hearts [31] using various injury models. Due to a lack of lineage tracing transgenic tools in these organisms, the source of new myocardium has not been definitively shown. However, Braun and colleagues showed a reduction in contractile protein expression after injury [32], reminiscent of the cardiomyocyte dedifferentiation observed in zebrafish heart regeneration [16, 17], suggesting a possible common mechanism. Not surprisingly, changes in extracellular matrix protein expression were also shown to accompany adult newt heart regeneration. Of particular interest, tenascin C was found to increase newt cardiomyocyte cell cycle re-entry in vitro [33]. However, evidence for cytokinesis was not shown. Interestingly, matrix production and remodeling enzymes were shown to change along with differentiation of immortalized CPCs in vitro, providing a direct link between the state of cardiomyocyte maturation and extracellular matrix remodeling [34].

Some reports have suggested that accelerated lower vertebrate regeneration is a consequence of cellular plasticity. For example, adult newt cardiomyocytes have been shown to transdifferentiate toward skeletal myocyte or chondrocyte lineages after transplantation into regenerating limb blastema [32]. Conversely, transdifferentiation was not observed during in vitro culture or after transplantation into intact limbs. It would be interesting to see if adult mammalian cardiomyocytes can be transdifferentiated by amphibian blastema; this would indicate a conserved intrinsic regenerative program within vertebrate cardiomyocytes and a non-conserved extrinsic tissue response to injury.

Although adult mammalian hearts do not efficiently regenerate, Olson and colleagues showed in 2011 that neonatal mice (up to postnatal day 7) can regenerate their heart after apical resection [35]. Genetic lineage tracing experiments showed that, similar to zebrafish, the cardiomyocytes are repopulated by pre-existing cardiomyocytes. Immunostaining with anti-Troponin antibodies demonstrated sarcomeric disassembly in myocytes, again suggesting dedifferentiation and expansion of resident cardiomyocytes as a driver of regeneration. Notably, there has been some controversy over the extent of neonatal cardiac regeneration, where it has been suggested that neonatal hearts heal by scarring after apical resection [36]. However, several investigators report the reproducibility of neonatal heart regeneration in an apical resection model and have suggested technical differences as a source of variability [37]. Furthermore, it is not surprising that the severity of injury influences the efficiency of regeneration [38].

Whether or not neonatal hearts exhibit complete regeneration in response to injury, their apparent neomyogenic capacity is a major point of focus that could potentially be used clinically if similar mechanisms can be exploited in the adult myocardium. Thus, it is important to critically evaluate not only the functional recovery after MI, but also the extent of new cardiomyocyte generation in neonatal mice. To that end, cell cycle re-entry of neonatal cardiomyocytes has been thoroughly demonstrated. Soonpaa et al. used tritiated thymidine to demonstrate a spike in S-phase DNA synthesis in neonatal murine cardiomyocytes, beginning near birth and persisting throughout the first week of life [39]. The fraction of binucleated cardiomyocytes increased steadily during this period as the cells lost the ability to complete cytokinesis.

In contrast to S-phase re-entry, the study of cell division is currently more technically challenging. Cytokinesis has traditionally been evaluated using antibodies against cleavage furrow markers such as Aurora B kinase. These techniques can be difficult to interpret with in vivo or in vitro samples, since staining in closely associated non-cardiomyocytes could contribute to false-positive results. This has led investigators to explore alternative methods, such as mosaic analysis with double markers (MADM), to genetically trace divided cardiomyocytes [40]. Interestingly, pulsing of MADM transgenic mice with tamoxifen between postnatal day 2 and 8 revealed that 5 % of labeled MYH6-expressing cardiomyocytes had undergone cytokinesis, giving rise to single labeled (GFP^+ or RFP^+) cells. Due to differential sorting of chromosomes, as well as non-sortable labeling in G0/G1, this figure likely underestimates the actual rate of cytokinesis in labeled cardiomyocytes. Furthermore, it is unclear whether Cre-mediated interchromosomal recombination is unbiased with respect to different cellular states in the heterogeneous cardiomyocyte population. Thus, at this time it is difficult to quantify the actual rate of cardiomyocyte cell division. Nonetheless, it is generally accepted that a significant proportion of neonatal cardiomyocytes have the ability complete cell division and contribute to cardiac regeneration. However, by postnatal day 7, murine cardiomyocytes have mostly exited the cell cycle [39] and lost their ability to regenerate injured myocardium [35].

Interestingly, it has been suggested that altered cardiac circulation accompanies newt heart regeneration, where blood is shunted away from the left ventricle [41]. This is reminiscent of enhanced cardiomyocyte cell cycle and myocardial remodeling in patients with ventricular assist device [42, 43], where a reduction in load may allow partial induction of a regenerative response. It would be interesting to see if neonatal mice exhibit a similar phenomenon during cardiac regeneration. For example, although functional closure of the ductus arteriosus occurs within 3 h post-birth in mice, remodeling takes place over several weeks [41]. Thus, additional studies would be prudent to evaluate the possibility of compensatory shunting of circulation during ventricular regeneration in neonatal mice.

6.1.3 Developments in Induced Heart Regeneration

Despite significant progress in understanding regenerative processes in lower vertebrates and in neonatal mice, it is still unclear how many of these findings can be applied to induce cardiac regeneration in adult mammals. The observation that neonatal mouse hearts can regenerate cardiac injuries is alluring, but there are major differences between neonates and adults with respect to cardiac physiology at the cellular, tissue, and neurohumoral levels. A modest degree of cell cycle re-entry has been observed in adult human and mouse cardiomyocytes [39, 44–46], but evidence for cardiomyocyte cell division in adult mammals is scant. To estimate human cardiomyocyte turnover, Bergmann et al. took advantage of a period of nuclear bomb testing in the 1950s and 1960s, which resulted in a pulse of atmospheric ^{14}C

eventually being incorporated into newly synthesized DNA in human cardiomyocytes [44, 45]. They found that less than 1 % cardiomyocytes were turned over annually in adult humans. Additionally, they showed that DNA content increased in the first 10 years of human life, until most cardiomyocytes were tetraploid [44]. In contrast to mice, most adult human cardiomyocytes are mononucleate [47]. Together, these results indicate that most human cardiomyocytes terminally exit the cell cycle before karyokinesis, whereas mouse cardiomyocytes tend to exit the cell cycle after karyokinesis, but before cytokinesis [48].

Although measurement of cell division in human cardiomyocytes is extremely difficult, recent advances in lineage tracing technology have enabled definitive labeling of divided cardiomyocytes in mice. A recent study using mosaic analysis with double markers [49] showed that approximately 1 % of labeled adult cardiomyocytes had undergone cell division after 2 weeks of daily tamoxifen induction [40]. However, as discussed above, potential bias of interchromosomal recombination could obscure quantification of cell division. Importantly, myocardial infarction prior to labeling did not increase cell division, indicating a lack of regeneration in adult mouse hearts. Still, the immense burden on human health has warranted an abundance of investigations seeking the ultimate feat of cardiovascular medicine: induced adult human heart regeneration.

Numerous strategies have been devised to induce adult mammalian heart regeneration and typically rely on mouse models of myocardial infarction, such as permanent left anterior descending (LAD) artery ligation [50, 51]. Ischemia-reperfusion (IR) models [52] are an even better representation of human myocardial infarction, due to post-MI surgical intervention [8, 9]. Large animal models [53, 54] are useful to translate findings in mice and to test regenerative strategies that are difficult in rodent models due to differences in anatomy, physiology or scalability.

Here, we discuss various therapeutic approaches (summarized in Fig. 6.1) to induce mammalian heart regeneration, including strategies that augment endogenous cardiac regeneration, or supply an exogenous source of cardiomyocyte replacement, consisting of allografts or the re-introduction of modified autologous cells.

6.1.4 Cardiac Progenitor Cells

Attempts to stimulate endogenous heart regeneration and replenish lost cardiomyocytes has been in part motivated by the hypothetical existence of a population of resident or non-resident cardiac progenitor cells (CPCs), which were thought to be a renewable source of committed cardiomyogenic cells. In theory, either autologous or allogeneic CPCs could conceivably be grafted into ischemic injuries to facilitate cardiac regeneration. However, several supposed CPC cell types have ultimately been found to represent at best a very rare contributor to new cardiomyocytes in vivo. For example, *Lin*$^{-}$*c-kit*$^{+}$ CPCs initially showed promise for adult mammalian heart regeneration [55]. However, these cells were later reported to have limited utility in induced adult mammalian heart regeneration, despite their potential to support regeneration in

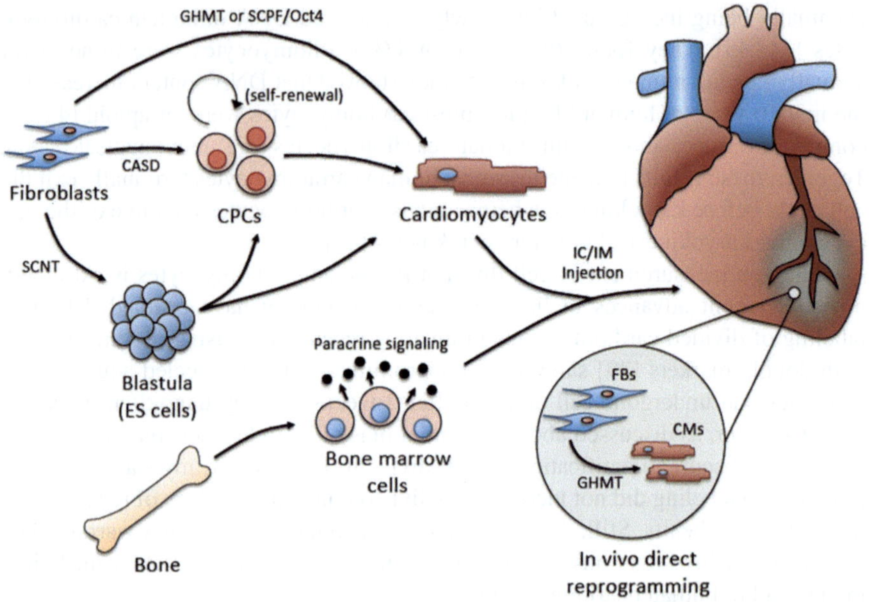

Fig. 6.1 Autologous cellular approaches to cardiac regeneration. Promising sources of autologous patient cells for therapeutic cardiac regeneration include dermal fibroblasts and bone marrow cells, which can be delivered to the infarct via intracoronary (IC) or intramyocardial (IM) injection. Bone marrow cells are thought to act via paracrine effects to encourage regeneration. Fibroblasts can be converted directly to cardiomyocyte-like cells via GHMT or small molecules (SCPF) and Oct4. An expandable population of cardiac progenitors can be created using cell activated and signaling-directed (CASD) lineage conversion. CPCs and cardiomyocytes can also be created via embryonic stem cells created using somatic cell nuclear transfer (SCNT). (*Inset*) In vivo reprogramming can be used to convert resident cardiac fibroblasts into cardiomyocyte-like cells in situ using GHMT factors

neonates [56, 57]. A recent article confirmed the lack of significant direct contribution by cardiac resident c-kit$^+$ progenitors to new cardiomyocytes [58]. Specifically, *c-kit*$^+$ cells did not co-express *Nkx2.5* or sarcomeric proteins at any stage, but were consistently found to co-express the endothelial marker CD31. Furthermore, endothelial-specific *Tie2*-driven expression of Cre completely abolished a *c-kit* driven floxed LacZ reporter. Thus, despite the observation of c-kit$^+$ cells in both the developing and adult heart, they were found to contribute mostly to endothelial cells, rather than cardiomyocytes. As an exogenous cell therapy for heart regeneration [59], it seems likely that any potential benefit of *c-kit*$^+$ progenitor cells to cardiac function would be indirect, for example through paracrine signaling. Other potential endogenous adult murine CPCs have been described, such as *Sca1*$^+$ cells [60, 61]. However ectopic Cre-expression may have confounded initial interpretations of *Sca1*$^+$ CPCs, and the lack of a human ortholog limits the application to human heart failure therapy (reviewed in [62]).

By contrast, *Isl1*$^+$ cells are a true cardiomyocyte progenitor population derived from the second heart field and have been shown to give rise to a majority of cardiomyocytes in the developing mouse heart [63, 64]. Cre-based lineage tracing experiments showed that by embryonic day 9.75, *Isl1*$^+$ progenitor cells generated nearly all cells in the outflow tract and right ventricle, as well as 65 % of the left atria and 20 % of the left ventricle [63]. Moretti et al. showed that *Isl1*$^+$ precursors are multipotent and could give rise to smooth muscle and endothelial lineages in addition to cardiomyocytes [65]. They also demonstrated that *Isl1*$^+$ cells could be differentiated in vitro from ES cells and propagated on cardiac mesenchyme feeder layers, indicating a potential source of therapeutic progenitor cells for heart failure. A majority of the remaining heart, including the left ventricle, is derived from *Isl1*$^-$ progenitors from the primary heart field, characterized by expression of early developmental markers such as GATA4, NKX2.5, and TBX5 (reviewed in [66]).

The persistence of a clinically useful population of resident CPCs in adult mammalian hearts has been an elusive and ongoing pursuit. However, more tangible applications of developmental CPC research in heart regeneration have come through the use of CPC markers to identify potential alternative therapeutic cellular sources of neomyogenesis. Such induced CPCs can now be obtained by pretreatment of ES and iPS cells, as discussed below. Furthermore, the understanding of fetal heart development on the molecular level has led to the discovery of fetal gene re-expression during heart failure [12], which could represent failed attempts to regenerate the adult heart through developmental recapitulation.

6.1.5 Bone-Marrow Derived Cells

Bone marrow-derived cells (BMCs) represent an attractive source of regenerative therapy, since autologous donor tissue can be easily and safely obtained. Initial promise came from an early study that showed 5-azacytidine treatment could induce cardiomyocyte differentiation from immortalized BMCs in vitro [67]. Subsequently, it was shown that autologous BMCs could improve recovery after myocardial infarction in rats [68, 69]. A 2001 study showed a low rate of myocardial engraftment in an ischemia-reperfusion model after bone marrow transplantation of supposed multipotent *CD34*$^{-/low}$, *c-kit*$^+$, *sca1*$^+$ side population (SP) cells, obtained from *Rosa26-lacZ* donor mice [70]. The purity of the SP cells was high at 91 %, but a even a low rate of contamination by other cell types could confound the interpretation that SP cells themselves give rise to cardiomyocytes. Nevertheless, the observation that bone marrow derived cells could contribute to endothelial cells and cardiomyocytes at all was encouraging for future developments.

Numerous other pre-clinical and clinical studies have investigated the safety and efficacy of bone marrow-derived cell therapy on acute myocardial infarction and heart failure. Results from some individual clinical trials have been positive [71], but large-scale meta-analyses have shown either modest or no benefit on cardiac function or mortality [72, 73]. Looking forward, it will be interesting to see the results of an ongoing large scale phase III clinical trial testing the efficacy of intracoronary delivery of autologous BMCs [74].

6.1.6 Embryonic Stem Cells

Human embryonic stem (hES) cells can be obtained from sperm-fertilized blastocysts [75] or, more conveniently, produced from adult fibroblasts by somatic cell nuclear transfer into oocytes [76, 77]. Being pluripotent, ES cells have the ability to give rise to all three germ layers, including all cell types of the heart. Thus, ES cells are a promising source of cardiomyocyte replacement in the failing heart. However, teratoma formation from direct ES cell injection demonstrates that neither normal nor failing myocardium lacks the developmental signals for faithful differentiation into myocardial lineages [78, 79]. ES cell-derived cardiomyocytes (ES-CMs) can be differentiated from hES cells in vitro by treatment with activin A and BMP4 [80]. In an athymic rat IR model, it was shown that infarcted myocardium could be grafted with hES-CMs by direct cardiac injection [80]. Importantly, a pro-survival cocktail (containing cell adhesion promoting Matrigel, mitochondrial death inhibitors Bcl-KL peptide and cyclosporine A, vasodilator pinacidil, AKT activator IGF-1, and caspase inhibitor ZVAD-fmk) was used to improve graft survival and functional recovery.

Despite the initial excitement for ES-CM treatment, a later study showed that although both allogeneic undifferentiated ES cell and ES-CM treatment provided improvements to ejection fraction in infarcted mouse myocardium, the ES-CM treated groups had an increased risk of cardiac arrhythmia and death [81]. This observation was presumably due to incomplete maturity of *in vitro* differentiated hES-CMs, or alternatively to the mismatch in normal heart rate between human and mouse cardiomyocytes. A subsequent study using an immunocompromised guinea pig cryoinjury model showed engraftment by hES-derived cardiomyocytes with reduced arrhythmia [82]. However, a non-human primate model of the more relevant IR injury again showed significant arrhythmia after engraftment of hES-CMs [83].

These exciting developments in ES-derived myocardial grafts show promise for future heart failure treatments. However, there is a clear need to better understand cardiomyocyte differentiation and to develop protocols to create more mature cardiomyocyte grafts that can recapitulate native pacing. In that light, a recent study showed that 1 year old in vitro differentiated ES-CMs are more similar to mature myocardial tissue in vivo and that the let-7 miR family plays an important role in the maturation process [84]. Furthermore, an earlier study showed that forced expression of connexin 43 improved conduction not only in embryonic cardiomyocyte grafts, but even in skeletal myoblast grafts in infarcted mouse hearts [85].

Despite the use of ES cells as a powerful research tool, and the promising results of preclinical heart regeneration studies, reluctance to enter clinical trials hinges in part on their potential for immune rejection and tumorigenesis [86], not to mention ethical constraints. It will be interesting to see if future developments in autologous ES cell creation [76] and refinements in differentiation and purification protocols will change these perspectives.

6.1.7 Induced Pluripotent Stem Cells

In 2006, Takahashi and Yamanaka reported that adult fibroblasts could be reprogrammed to become induced pluripotent stem (iPS) cells [87]. By forced expression of *Oct3/4*, *Sox2*, *c-Myc*, and *Klf4*, adult mouse fibroblasts became competent for teratoma formation and differentiation into all three germ layers. However, it was still not clear whether the same protocol could be used with human cells. The following year, the same group reported that iPS cells could be generated using human fibroblasts [88]. This was a landmark development in regenerative medicine because it indicated that dispensable autologous adult donor tissue could be used to potentially regenerate any tissue, including the heart.

Although iPS cells theoretically should avoid complications due to immune rejection when using reprogrammed autologous cells, some evidence has suggested otherwise [89]. Furthermore, the tumorigenic risk of retrovirus-reprogrammed cells has led others to pursue chemical or protein-mediated derivation of reprogrammed cells [90, 91]. Still, the pluripotency of iPS cells necessitates a better understanding of differentiation and the development of robust progenitor purification before clinical applications can safely use iPS cells. Nonetheless, iPS cells have become an invaluable research tool and will continue to change the face of regenerative research.

6.1.8 Direct Reprogramming

The discovery of iPS cell reprogramming and the risk of teratoma/tumor formation from the use of pluripotent stem cells quickly led others to pursue alternative approaches to cellular reprogramming. Related approaches were then used to directly reprogram fibroblasts into induced cardiomyocyte-like (iCM) cells without a pluripotent intermediate. The motivation for this type of reprogramming lies in the abundance of fibroblasts in the infarcted myocardium that could serve as a source of new cardiomyocytes. A key observation that led to the discovery of direct reprogramming approaches was the recognition that several core transcription factors (GATA4, HAND2, MEF2C, MESP1, NKX2-5, and TBX5) play a major role in heart development and differentiation. In 2010, a subset of these factors, GMT (GATA4, MEF2C, and TBX5), was used to directly reprogram mouse cardiac and dermal fibroblasts into iCM cells in vitro [92]. Subsequently, in vivo reprogramming was achieved with either GMT or GHMT (GMT+HAND2), yielding improved cardiac function after myocardial infarction in mice [93, 94]. Co-injection of thymosin β4 with GMT reprogramming improved myocardial function after MI [93, 95]. Ding and colleagues showed that small molecules SCPF (SB431542, CHIR99021, parnate, and forskolin) and Oct4 alone could achieve direct reprogramming in vitro [96]. Alternative reprogramming formulations have since been developed, including a microRNA cocktail that effectively converts adult cardiac fibroblasts [97]. Importantly, Olson and colleagues reported a cardiac reprogramming cocktail that works in human cells [98]. Recently, it was shown that Akt1/protein kinase B enhances GHMT conversion efficiency and iCM maturity, including increased polynucleation [99].

In contrast to iPS cells, direct reprogramming offers a source of cardiomyocyte replacement that bypasses the teratoma-competent pluripotent stage. However, more efficient methods to convert and target cardiac fibroblasts need to be developed to move forward in the clinic [100]. In addition, the use of safe vectors or chemical approaches for reprogramming factors would expedite clinical utility of direct reprogramming [96, 100]. Furthermore, despite its promising direction, the tradeoff of reprogramming fibroblasts into cardiomyocytes must still be critically evaluated with respect to the loss of fibroblast function in the failing heart [101]. Perhaps the recent discovery of expandable induced cardiomyocyte-like progenitors [102] will lead to similar strategies that can address concerns of a fibroblast-cardiomycote tradeoff for in vivo conversion.

6.1.9 Dedifferentiated Adult Cardiomyocytes

Dedifferentiation of adult cardiomyocytes can be seen through the re-expression of fetal gene programs in heart failure [12]. Thus, it should not be surprising that adult mammalian cardiomyocytes can dedifferentiate to some degree in culture [103, 104]. Still, evidence for true adult cardiomyocyte cell division, even in the far-removed in vitro environment, is scarce. This suggests that despite varying degrees of dedifferentiation of adult cardiomyocytes in vitro and in vivo, there may exist an inherent block to actually complete cell division. This idea is further supported by the rarity of cardiomyocyte-derived cancers. Nevertheless, rare examples of significantly proliferating adult mammalian cardiomyocytes have been reported, such as rat cardiomyocytes showing high levels of bromodeoxyuridine (BrdU), Ki67 and phosphohistone 3 (PH3) staining in vitro [104]. Recently, the dedifferentiation process of these cultured myocytes was shown to be regulated by epigenomic reprogramming [105].

Fascinatingly, explanted cardiac tissue, cultured under non-adhesive conditions, has been shown to recapitulate a stem cell-like niche that apparently contributes to myocardial repair [106]. The cell preparations derived from such cultures, deemed cardiosphere-derived cells (CDCs) are now being evaluated for the treatment of heart failure in humans. Phase I clinical trials have shown positive results with an increase in viable mass and a reduction in scar size [107, 108]. Interestingly, it was recently shown that exosomes from CDCs may help mediate their regenerative effects [109]. It will be interesting to see how ongoing clinical trials could potentially improve patient outcome [110].

6.1.10 Stimulation of Adult Cardiomyocyte Proliferation

The induction of cardiomyocyte proliferation through cell cycle re-entry and true cell division has been a heavily sought goal of research, with the ultimate goal of adult human heart regeneration through the expansion and replenishment of endogenous cardiomyocytes. Numerous reports have demonstrated induced re-entry into S-phase by adult mammalian cardiomyocytes, for example by cell cycle activators Cyclin A2 [111] and E2F [112]. Although cytokinetic figures have been observed, robust cardiomyocyte cell division has been difficult to achieve. Immortalization with SV40TAg indicated that it is possible to induce persistent cell division in adult rat ventricular myocytes [113]. However, it is unclear what percentage of adult cardiomyocytes have the capacity to divide without apoptosis even under oncogenic conditions. Since the risk of tumorigenesis precludes serious consideration of SV40Tag in the clinic, the search for regulated stimulation of cardiomyocyte proliferation continues. Various approaches have since been used to increase cardiomyocyte proliferation and enhance MI repair, such as those involving miRNAs [114–116] and neuregulin [117, 118] signaling. The Hippo pathway has recently become an intense subject of investigation in heart regeneration due to its role in organ size control [119]. Modulation of the Hippo pathway has been shown to extend the developmental window of cardiomyocyte proliferation and offer modest improvements when administered after MI in several reports [120–122]. Despite promising results from many of these studies, the major cell cycle blocks in adult mammalian cardiomyocytes are largely not well understood. Furthermore, definitive regeneration in adult mammals is still an active pursuit with room for improvement.

6.1.11 Tissue Mechanics

As mentioned earlier, mechanical stiffness has been associated with reduced ventricular function and progressive heart failure. Recombinant elastin production by transduced endothelial cell transplants reduced infarct size and improved cardiac function after myocardial infarction in rats [123]. This result corroborates observations of progressive heart malfunction as a result of mechanically mediated myofibroblast conversion and runaway fibrosis accompanied by cardiomyocyte cell death (reviewed in [124]). Tissue mechanics has been shown to be important in several aspects of cardiomyocyte biology, such as contractility [125], development [126–128], differentiation [129], and maturation [130]. Recently, a collagen matrix patch containing FSTL1 was used to promote myocardial repair in a porcine myocardial infarction model [131]. It was found that therapeutic effect was influenced not only by the location of FSTL1 secretion, but also by the elasticity of the collagen patch. Thus, it is becoming increasingly clear that tissue/matrix mechanics plays an important role in cardiac disease and remodeling and should be carefully considered in future efforts to induce heart regeneration.

6.1.12 Engraftment

Engraftment of exogenous cells into the heart has been a challenging hurdle to treat heart disease via cellular approaches. The dynamic mechanical demands of the human heart, forcefully pumping at approximately 1 Hz, likely pose a thermodynamic barrier to cell attachment and integration within the dense extracellular matrix. Not surprisingly, there may be an age-dependence on the success of donor cell engraftment, as shown by higher engraftment of fetal and neonatal rat cardiomyocytes into injured and non-injured adult rat hearts when compared to adult cardiomyocyte engraftment [132]. Despite a higher rate of engraftment for younger donor tissue, engraftment cell survival is typically very low, even for stem and progenitor cell grafts [133]. Nevertheless, an enormous body of work describes various attempts to achieve therapeutic benefit from exogenous cell therapy in heart injury models, as reviewed above. Concurrent developments are underway to increase cell engraftment in the heart and other tissues, including cell adhesive matrices [134, 135] as well as cell pretreatment to increase cardiac homing (reviewed in [136]).

6.2 Conclusions

The field of regenerative biology has made enormous progress in understanding some of the species differences in cardiac regeneration and in the discovery of several therapeutic strategies that have shown some effect on mitigating the effects of human heart failure. However, the ultimate therapeutic endpoint is still out of reach, and further work will be required to obtain a better basic understanding of myocardial biology, including the molecular nature of adult cardiomyocyte cell cycle block, the role of tissue mechanics in heart disease, and the interplay between fibrosis and cardiomyocyte health. Exciting clinical and preclinical developments in cellular and molecular therapies utilizing cardiospheres or miRNA and Hippo signaling could be revealing in the oncoming years. Still, it will be crucial to continue the pursuit of basic discovery in cardiomyocyte biology and the refinement of drug, gene, and cell delivery approaches to maximize progress toward human heart regeneration.

References

1. Braunwald E (1997) Shattuck lecture-cardiovascular medicine at the turn of the millennium: triumphs, concerns, and opportunities. N Engl J Med 337(19):1360–1369. doi:10.1056/NEJM199711063371906
2. Askoxylakis V, Thieke C, Pleger ST et al (2010) Long-term survival of cancer patients compared to heart failure and stroke: a systematic review. BMC Cancer 10(1):105. doi:10.1186/1471-2407-10-105
3. Konstantinidis K, Whelan RS, Kitsis RN (2012) Mechanisms of cell death in heart disease. Arterioscler Thromb Vasc Biol 32(7):1552–1562. doi:10.1161/ATVBAHA.111.224915
4. Farb A, Kolodgie FD, Jenkins M et al (1993) Myocardial infarct extension during reperfusion after coronary artery occlusion: pathologic evidence. J Am Coll Cardiol 21(5):1245–1253. doi:10.1016/0735-1097(93)90253-W
5. Minezaki KK, Suleiman MS, Chapman RA (1994) Changes in mitochondrial function induced in isolated guinea-pig ventricular myocytes by calcium overload. J Physiol 476(3):459–471

6. Ferrari R, Alfieri O, Curello S et al (1990) Occurrence of oxidative stress during reperfusion of the human heart. Circulation 81(1):201–211. doi:10.1161/01.CIR.81.1.201
7. Opie LH (1991) Role of calcium and other ions in reperfusion injury. Cardiovasc Drugs Ther 5(Suppl 2):237–247
8. Beyersdorf F (2009) The use of controlled reperfusion strategies in cardiac surgery to minimize ischaemia/reperfusion damage. Cardiovasc Res 83(2):262–268, 10.1093/cvr/cvp110
9. Hausenloy DJ, Yellon DM (2013) Myocardial ischemia-reperfusion injury: a neglected therapeutic target. J Clin Invest 123(1):92–100. doi:10.1172/JCI62874
10. Weber KT, Sun Y, Bhattacharya SK, Ahokas RA, Gerling IC (2013) Myofibroblast-mediated mechanisms of pathological remodelling of the heart. Nat Rev Cardiol 10(1):15–26. doi:10.1038/nrcardio.2012.158
11. Rysä J, Leskinen H, Ilves M, Ruskoaho H (2005) Distinct upregulation of extracellular matrix genes in transition from hypertrophy to hypertensive heart failure. Hypertension 45(5):927–933, 10.1093/cvr/cvp110
12. Dirkx E, da Costa Martins PA, De Windt LJ (2013) Regulation of fetal gene expression in heart failure. Biochim Biophys Acta 1832(12):2414–2424. doi:10.1016/j.bbadis.2013.07.023
13. Kolwicz SC, Purohit S, Tian R (2013) Cardiac metabolism and its interactions with contraction, growth, and survival of cardiomyocytes. Circ Res 113(5):603–616. doi:10.1161/CIRCRESAHA.113.302095
14. Jaźwińska A, Sallin P (2016) Regeneration versus scarring in vertebrate appendages and heart. J Pathol 238(2):233–246. doi:10.1002/path.4644
15. Judd J, Xuan W, Huang GN (2015) Cellular and molecular basis of cardiac regeneration. Turk J Biol 40(2):265–275
16. Kikuchi K, Holdway JE, Werdich AA et al (2010) Primary contribution to zebrafish heart regeneration by gata4(+) cardiomyocytes. Nature 464(7288):601–605. doi:10.1038/nature08804
17. Jopling C, Sleep E, Raya M et al (2010) Zebrafish heart regeneration occurs by cardiomyocyte dedifferentiation and proliferation. Nature 464(7288):606–609. doi:10.1038/nature08899
18. Sallin P, de Preux Charles A-S, Duruz V, Pfefferli C, Jaźwińska A (2015) A dual epimorphic and compensatory mode of heart regeneration in zebrafish. Dev Biol 399(1):27–40. doi:10.1016/j.ydbio.2014.12.002
19. Kikuchi K, Holdway JE, Major RJ et al (2011) Retinoic acid production by endocardium and epicardium is an injury response essential for zebrafish heart regeneration. Dev Cell 20(3):397–404. doi:10.1016/j.devcel.2011.01.010
20. Lepilina A, Coon AN, Kikuchi K et al (2006) A dynamic epicardial injury response supports progenitor cell activity during zebrafish heart regeneration. Cell 127(3):607–619, 10.1016/j.cell.2006.08.052
21. Wang J, Karra R, Dickson AL, Poss KD (2013) Fibronectin is deposited by injury-activated epicardial cells and is necessary for zebrafish heart regeneration. Dev Biol 382(2):427–435. doi:10.1016/j.ydbio.2013.08.012
22. Willems IE, Arends JW, Daemen MJ (1996) Tenascin and fibronectin expression in healing human myocardial scars. J Pathol 179(3):321–325. doi:10.1007/s12265-012-9406-3
23. Knowlton AA, Connelly CM, Romo GM et al (1992) Rapid expression of fibronectin in the rabbit heart after myocardial infarction with and without reperfusion. J Clin Invest 89(4):1060–1068. doi:10.1172/JCI115685
24. Altrock E, Sens C, Wuerfel C et al (2015) Inhibition of fibronectin deposition improves experimental liver fibrosis. J Hepatol 62(3):625–633
25. Serini G, Bochaton-Piallat M-L, Ropraz P et al (1998) The fibronectin domain ED-A is crucial for myofibroblastic phenotype induction by transforming growth factor-β1. J Cell Biol 142(3):873–881. doi:10.1016/j.jhep.2014.06.010
26. Bhattacharyya S, Tamaki Z, Wang W et al (2014) FibronectinEDA promotes chronic cutaneous fibrosis through Toll-like receptor signaling. Sci Transl Med 6(232):232ra50. doi:10.1126/scitranslmed.3008264

27. Bettencourt-Dias M, Mittnacht S, Brockes JP (2003) Heterogeneous proliferative potential in regenerative adult newt cardiomyocytes. J Cell Sci 116(Pt 19):4001–4009. doi:10.1242/jcs.00698
28. Witman N, Murtuza B, Davis B, Arner A, Morrison JI (2011) Recapitulation of developmental cardiogenesis governs the morphological and functional regeneration of adult newt hearts following injury. Dev Biol 354(1):67–76. doi:10.1016/j.ydbio.2011.03.021
29. Oberpriller JO, Oberpriller JC (1974) Response of the adult newt ventricle to injury. J Exp Zool 187(2):249–253
30. Piatkowski T, Mühlfeld C, Borchardt T, Braun T (2013) Reconstitution of the myocardium in regenerating newt hearts is preceded by transient deposition of extracellular matrix components. Stem Cells Dev 22(13):1921–1931. doi:10.1089/scd.2012.0575
31. Cano-Martínez A, Vargas-González A, Guarner-Lans V et al (2010) Functional and structural regeneration in the axolotl heart (*Ambystoma mexicanum*) after partial ventricular amputation. Arch Cardiol Mex 80(2):79–86
32. Laube F, Heister M, Scholz C, Borchardt T, Braun T (2006) Re-programming of newt cardiomyocytes is induced by tissue regeneration. J Cell Sci 119(Pt 22):4719–4729. doi:10.1242/jcs.03252
33. Mercer SE, Odelberg SJ, Simon H-G (2013) A dynamic spatiotemporal extracellular matrix facilitates epicardial-mediated vertebrate heart regeneration. Dev Biol 382(2):457–469. doi:10.1016/j.ydbio.2013.08.002
34. Bax NAM, van Marion MH, Shah B et al (2012) Matrix production and remodeling capacity of cardiomyocyte progenitor cells during in vitro differentiation. J Mol Cell Cardiol 53(4):497–508. doi:10.1016/j.yjmcc.2012.07.003
35. Porrello ER, Mahmoud AI, Simpson E et al (2011) Transient regenerative potential of the neonatal mouse heart. Science 331(6020):1078–1080. doi:10.1126/science.1200708
36. Andersen DC, Ganesalingam S, Jensen CH, Sheikh SP (2014) Do neonatal mouse hearts regenerate following heart apex resection? Stem Cell Reports 2(4):406–413. doi:10.1016/j.stemcr.2014.02.008
37. Sadek HA, Martin JF, Takeuchi JK et al (2014) Multi-investigator letter on reproducibility of neonatal heart regeneration following apical resection. Stem Cell Reports 3(1):1. doi:10.1016/j.stemcr.2014.06.009
38. Bryant DM, O'Meara CC, Ho NN et al (2015) A systematic analysis of neonatal mouse heart regeneration after apical resection. J Mol Cell Cardiol 79:315–918. doi:10.1016/j.yjmcc.2014.12.011
39. Soonpaa MH, Kim KK, Pajak L, Franklin M, Field LJ (1996) Cardiomyocyte DNA synthesis and binucleation during murine development. Am J Physiol 271(5 Pt 2):H2183–H2189
40. Ali SR, Hippenmeyer S, Saadat LV et al (2014) Existing cardiomyocytes generate cardiomyocytes at a low rate after birth in mice. Proc Natl Acad Sci U S A 111(24):8850–8855. doi:10.1073/pnas.1408233111
41. Tada T, Kishimoto H (1990) Ultrastructural and histological studies on closure of the mouse ductus arteriosus. Acta Anat 139(4):326–334
42. Canseco DC, Kimura W, Garg S et al (2015) Human ventricular unloading induces cardiomyocyte proliferation. J Am Coll Cardiol 65(9):892–900. doi:10.1016/j.jacc.2014.12.027
43. Dandel M, Weng Y, Siniawski H et al (2011) Heart failure reversal by ventricular unloading in patients with chronic cardiomyopathy: criteria for weaning from ventricular assist devices. Eur Heart J 32(9):1148–1160. doi:10.1093/eurheartj/ehq353
44. Bergmann O, Bhardwaj RD, Bernard S et al (2009) Evidence for cardiomyocyte renewal in humans. Science 324(5923):98–102. doi:10.1126/science.1164680
45. Bergmann O, Zdunek S, Felker A et al (2015) Dynamics of cell generation and turnover in the human heart. Cell 161(7):1566–1575. doi:10.1016/j.cell.2015.05.026
46. Soonpaa MH, Field LJ (1994) Assessment of cardiomyocyte DNA synthesis during hypertrophy in adult mice. Am J Physiol 266(4 Pt 2):H1439–H1445
47. Olivetti G, Cigola E, Maestri R et al (1996) Aging, cardiac hypertrophy and ischemic cardiomyopathy do not affect the proportion of mononucleated and multinucleated myocytes in the human heart. J Mol Cell Cardiol 28(7):1463–1477. doi:10.1006/jmcc.1996.0137

48. Engel FB, Schebesta M, Keating MT (2006) Anillin localization defect in cardiomyocyte binucleation. J Mol Cell Cardiol 41(4):601–612. doi:10.1016/j.yjmcc.2006.06.012
49. Zong H, Espinosa JS, Su HH, Muzumdar MD, Luo L (2005) Mosaic analysis with double markers in mice. Cell 121(3):479–492. doi:10.1016/j.cell.2005.0
50. Mahmoud AI, Porrello ER, Kimura W, Olson EN, Sadek HA (2009) Surgical models for cardiac regeneration in neonatal mice. Nat Protoc 9(2):305–311. doi:10.1038/nprot.2014.021
51. Curaj A, Simsekyilmaz S, Staudt M, Liehn E (2015) Minimal invasive surgical procedure of inducing myocardial infarction in mice. J Vis Exp 99, e52197. doi:10.3791/52197
52. Xu Z, Alloush J, Beck E, Weisleder N (2015) A murine model of myocardial ischemia-reperfusion injury through ligation of the left anterior descending artery. J Vis Exp 86, e51329. doi:10.3791/51329
53. Chong JJH, Murry CE (2014) Cardiac regeneration using pluripotent stem cells–progression to large animal models. Stem Cell Res 13(3 Pt B):654–665. doi:10.1016/j.scr.2014.06.005
54. Dixon JA, Spinale FG (2009) Large animal models of heart failure: a critical link in the translation of basic science to clinical practice. Circ Heart Fail 2(3):262–271. doi:10.1161/CIRCHEARTFAILURE.108.814459
55. Beltrami AP, Barlucchi L, Torella D et al (2003) Adult cardiac stem cells are multipotent and support myocardial regeneration. Cell 114(6):763–776. doi:10.1016/S0092-8674(03)00687-1
56. Zaruba M-M, Soonpaa M, Reuter S, Field LJ (2010) Cardiomyogenic potential of C-kit(+)-expressing cells derived from neonatal and adult mouse hearts. Circulation 121(18):1992–2000. doi:10.1161/CIRCULATIONAHA.109.909093
57. Jesty SA, Steffey MA, Lee FK et al (2012) C-kit+precursors support postinfarction myogenesis in the neonatal, but not adult, heart. Proc Natl Acad Sci U S A 109(33):13380–13385. doi:10.1073/pnas.1208114109
58. Sultana N, Zhang L, Yan J et al (2015) Resident c-kit(+) cells in the heart are not cardiac stem cells. Nat Commun 6:8701. doi:10.1038/ncomms9701
59. Bolli R, Chugh AR, D'Amario D et al (2011) Cardiac stem cells in patients with ischaemic cardiomyopathy (SCIPIO): initial results of a randomised phase 1 trial. Lancet 378(9806):1847–1857. doi:10.1016/S0140-6736(11)61590-0
60. Oh H, Bradfute SB, Gallardo TD et al (2003) Cardiac progenitor cells from adult myocardium: homing, differentiation, and fusion after infarction. Proc Natl Acad Sci U S A 100(21):12313–12318. doi:10.1073/pnas.2132126100
61. Uchida S, De Gaspari P, Kostin S et al (2013) Sca1-derived cells are a source of myocardial renewal in the murine adult heart. Stem Cell Reports 1(5):397–410. doi:10.1016/j.stemcr.2013.09.004
62. van Berlo JH, Molkentin JD (2014) An emerging consensus on cardiac regeneration. Nat Med 20(12):1386–1393. doi:10.1038/nm.3764
63. Cai C-L, Liang X, Shi Y et al (2003) Isl1 identifies a cardiac progenitor population that proliferates prior to differentiation and contributes a majority of cells to the heart. Dev Cell 5(6):877–889. doi:10.1016/S1534-5807(03)00363-0
64. Laugwitz K-L, Moretti A, Lam J et al (2005) Postnatal isl1+ cardioblasts enter fully differentiated cardiomyocyte lineages. Nature 433(7026):647–653. doi:10.1038/nature03215
65. Moretti A, Caron L, Nakano A et al (2006) Multipotent embryonic isl1+ progenitor cells lead to cardiac, smooth muscle, and endothelial cell diversification. Cell 127(6):1151–1165. doi:10.1016/j.cell.2006.10.029
66. Xin M, Olson EN, Bassel-Duby R (2013) Mending broken hearts: cardiac development as a basis for adult heart regeneration and repair. Nat Rev Mol Cell Biol 14(8):529–541. doi:10.1038/nrm3619
67. Makino S, Fukuda K, Miyoshi S et al (1999) Cardiomyocytes can be generated from marrow stromal cells in vitro. J Clin Invest 103(5):697–705
68. Tomita S, Li RK, Weisel RD et al (1999) Autologous transplantation of bone marrow cells improves damaged heart function. Circulation 100(Suppl II):247–256. doi:10.1161/01.CIR.100.suppl_2.II-247

69. Kocher AA, Schuster MD, Szabolcs MJ et al (2001) Neovascularization of ischemic myocardium by human bone-marrow-derived angioblasts prevents cardiomyocyte apoptosis, reduces remodeling and improves cardiac function. Nat Med 7(4):430–436. doi:10.1038/86498
70. Jackson KA, Majka SM, Wang H et al (2001) Regeneration of ischemic cardiac muscle and vascular endothelium by adult stem cells. J Clin Invest 107(11):1395–1402
71. Mathiasen AB, Qayyum AA, Jørgensen E et al (2015) Bone marrow-derived mesenchymal stromal cell treatment in patients with severe ischaemic heart failure: a randomized placebo-controlled trial (MSC-HF trial). Eur Heart J 36(27):1744–1753. doi:10.1093/eurheartj/ehv136
72. Nowbar AN, Mielewczik M, Karavassilis M et al (2014) Discrepancies in autologous bone marrow stem cell trials and enhancement of ejection fraction (DAMASCENE): weighted regression and meta-analysis. Br Med J 348:g2688. doi:10.1136/bmj.g2688
73. Gyöngyösi M, Wojakowski W, Lemarchand P et al (2015) Meta-Analysis of Cell-based CaRdiac stUdiEs (ACCRUE) in patients with acute myocardial infarction based on individual patient data. Circ Res 116(8):1346–1360. doi:10.1161/CIRCRESAHA.116.304346
74. Mathur A. BAMI. The effect of intracoronary reinfusion of Bone Marrow-derived Mononuclear Cells (BM-MNC) on all cause mortality in acute myocardial infarction. https://clinicaltrials.gov/ct2/show/nct01569178. Accessed 5 Apr 2016
75. Thomson JA (1998) Embryonic stem cell lines derived from human blastocysts. Science 282(5391):1145–1147. doi:10.1126/science.282.5391.1145
76. Tachibana M, Amato P, Sparman M et al (2013) Human embryonic stem cells derived by somatic cell nuclear transfer. Cell 154(2):1228–1238. doi:10.1016/j.cell.2013.05.006
77. Chung YG, Eum JH, Lee JE et al (2014) Human somatic cell nuclear transfer using adult cells. Cell Stem Cell 14(6):777–780. doi:10.1016/j.stem.2014.03.015
78. Nussbaum J, Minami E, Laflamme MA et al (2007) Transplantation of undifferentiated murine embryonic stem cells in the heart: teratoma formation and immune response. FASEB J 21(7):1345–1357. doi:10.1096/fj.06-6769com
79. Swijnenburg R-J, Tanaka M, Vogel H et al (2005) Embryonic stem cell immunogenicity increases upon differentiation after transplantation into ischemic myocardium. Circulation 112(Suppl 9):I166–I172. doi:10.1161/CIRCULATIONAHA.104.525824
80. Laflamme MA, Chen KY, Naumova AV et al (2007) Cardiomyocytes derived from human embryonic stem cells in pro-survival factors enhance function of infarcted rat hearts. Nat Biotechnol 25(9):1015–1024. doi:10.1038/nbt1327
81. Liao S-Y, Liu Y, Siu C-W et al (2010) Proarrhythmic risk of embryonic stem cell-derived cardiomyocyte transplantation in infarcted myocardium. Heart Rhythm 7(12):1852–1859. doi:10.1016/j.hrthm.2010.09.006
82. Shiba Y, Fernandes S, Zhu W-Z et al (2012) Human ES-cell-derived cardiomyocytes electrically couple and suppress arrhythmias in injured hearts. Nature 489(7415):322–325. doi:10.1038/nature11317
83. Chong JJH, Yang X, Don CW et al (2014) Human embryonic-stem-cell-derived cardiomyocytes regenerate non-human primate hearts. Nature 510(7504):273–277. doi:10.1038/nature13233
84. Kuppusamy KT, Jones DC, Sperber H et al (2015) Let-7 family of microRNA is required for maturation and adult-like metabolism in stem cell-derived cardiomyocytes. Proc Natl Acad Sci U S A 112(21):E2785–E2794. doi:10.1073/pnas.1424042112
85. Roell W, Lewalter T, Sasse P et al (2007) Engraftment of connexin 43-expressing cells prevents post-infarct arrhythmia. Nature 450(7171):819–824. doi:10.1038/nature06321
86. Sanganalmath SK, Bolli R (2013) Cell therapy for heart failure: a comprehensive overview of experimental and clinical studies, current challenges, and future directions. Circ Res 113(6):810–834. doi:10.1161/CIRCRESAHA.113.300219
87. Takahashi K, Yamanaka S (2006) Induction of pluripotent stem cells from mouse embryonic and adult fibroblast cultures by defined factors. Cell 126(4):663–676. doi:10.1016/j.cell.2006.07.024
88. Takahashi K, Tanabe K, Ohnuki M et al (2007) Induction of pluripotent stem cells from adult human fibroblasts by defined factors. Cell 131(5):861–872. doi:10.1016/j.cell.2007.11.019

89. Zhao T, Zhang Z-N, Rong Z, Xu Y (2011) Immunogenicity of induced pluripotent stem cells. Nature 474(7350):212–215. doi:10.1038/nature10135
90. Xie M, Cao N, Ding S (2014) Small molecules for cell reprogramming and heart repair: progress and perspective. ACS Chem Biol 9(1):34–44. doi:10.1021/cb400865w
91. Zhou H, Wu S, Joo JY et al (2009) Generation of induced pluripotent stem cells using recombinant proteins. Cell Stem Cell 4(5):381–384. doi:10.1016/j.stem.2009.04.005
92. Ieda M, Tsuchihashi T, Ivey KN et al (2009) Cardiac fibroblasts regulate myocardial proliferation through beta1 integrin signaling. Dev Cell 16(2):233–244. doi:10.1016/j.devcel.2008.12.007
93. Qian L, Huang Y, Spencer CI et al (2012) In vivo reprogramming of murine cardiac fibroblasts into induced cardiomyocytes. Nature 485(7400):593–598. doi:10.1038/nature11044
94. Song K, Nam Y-J, Luo X et al (2012) Heart repair by reprogramming non-myocytes with cardiac transcription factors. Nature 485(7400):599–604. doi:10.1038/nature11139
95. Srivastava D, Ieda M, Fu J, Qian L (2012) Cardiac repair with thymosin β4 and cardiac reprogramming factors. Ann N Y Acad Sci 1270:66–72. doi:10.1111/j.1749-6632.2012.06696.x
96. Wang H, Cao N, Spencer CI et al (2014) Small molecules enable cardiac reprogramming of mouse fibroblasts with a single factor, Oct4. Cell Rep 6(5):951–960. doi:10.1016/j.celrep.2014.01.038
97. Jayawardena TM, Egemnazarov B, Finch EA et al (2012) MicroRNA-mediated in vitro and in vivo direct reprogramming of cardiac fibroblasts to cardiomyocytes. Circ Res 110(11):1465–1473. doi:10.1161/CIRCRESAHA.112.269035
98. Nam Y-J, Song K, Luo X et al (2013) Reprogramming of human fibroblasts toward a cardiac fate. Proc Natl Acad Sci U S A 110(14):5588–5593. doi:10.1073/pnas.1301019110
99. Zhou H, Dickson ME, Kim MS, Bassel-Duby R, Olson EN (2015) Akt1/protein kinase B enhances transcriptional reprogramming of fibroblasts to functional cardiomyocytes. Proc Natl Acad Sci U S A 112(38):11864–11869. doi:10.1073/pnas.1516237112
100. Zhao Y, Londono P, Cao Y et al (2015) High-efficiency reprogramming of fibroblasts into cardiomyocytes requires suppression of pro-fibrotic signalling. Nat Commun 6:8243. doi:10.1038/ncomms9243
101. Nagalingam RS, Safi HA, Czubryt MP (2015) Gaining myocytes or losing fibroblasts: challenges in cardiac fibroblast reprogramming for infarct repair. J Mol Cell Cardiol 93:108–114. doi:10.1016/j.yjmcc.2015.11.029
102. Zhang Y, Cao N, Huang Y et al (2016) Expandable cardiovascular progenitor cells reprogrammed from fibroblasts. Cell Stem Cell 18(3):368–381. doi:10.1016/j.stem.2016.02.010
103. Nag AC, Cheng M (1986) Biochemical evidence for cellular dedifferentiation in adult rat cardiac muscle cells in culture: expression of myosin isozymes. Biochem Biophys Res Commun 137(2):855–862
104. Zhang Y, Li T-S, Lee S-T et al (2010) Dedifferentiation and proliferation of mammalian cardiomyocytes. PLoS One 5(9), e12559. doi:10.1371/journal.pone.0012559
105. Zhang Y, Zhong JF, Qiu H et al (2015) Epigenomic reprogramming of adult cardiomyocyte-derived cardiac progenitor cells. Sci Rep 5:17686. doi:10.1038/srep17686
106. Li T-S, Cheng K, Lee S-T et al (2010) Cardiospheres recapitulate a niche-like microenvironment rich in stemness and cell-matrix interactions, rationalizing their enhanced functional potency for myocardial repair. Stem Cells 28(11):2088–2098. doi:10.1002/stem.532
107. Malliaras K, Makkar RR, Smith RR et al (2014) Intracoronary cardiosphere-derived cells after myocardial infarction: evidence of therapeutic regeneration in the final 1-year results of the CADUCEUS trial (CArdiosphere-Derived aUtologous stem CElls to reverse ventricUlar dySfunction). J Am Coll Cardiol 63(2):110–122. doi:10.1016/j.jacc.2013.08.724
108. Makkar RR, Smith RR, Cheng K et al (2012) Intracoronary cardiosphere-derived cells for heart regeneration after myocardial infarction (CADUCEUS): a prospective, randomised phase 1 trial. Lancet 379(9819):895–904. doi:10.1016/S0140-6736(12)60195-0
109. Ibrahim AG-E, Cheng K, Marbán E (2014) Exosomes as critical agents of cardiac regeneration triggered by cell therapy. Stem Cell Reports 2(5):606–619. doi:10.1016/j.stemcr.2014.04.006

110. Allogeneic heart stem cells to achieve myocardial regeneration—full text view. [Internet]. Available from: https://clinicaltrials.gov/ct2/show/NCT01458405. Accessed 6 Apr 2016
111. Chaudhry HW, Dashoush NH, Tang H et al (2004) Cyclin A2 mediates cardiomyocyte mitosis in the postmitotic myocardium. J Biol Chem 279(34):35858–35866. doi:10.1074/jbc.M404975200
112. Ebelt H, Hufnagel N, Neuhaus P et al (2005) Divergent siblings: E2F2 and E2F4 but not E2F1 and E2F3 induce DNA synthesis in cardiomyocytes without activation of apoptosis. Circ Res 96(5):509–517. doi:10.1161/01.RES.0000159705.17322.57
113. Miller C, Rulfs J, Jaspers SR, Buckholt M, Miller TB (1994) Transformation of adult ventricular myocytes with the temperature sensitive A58 (tsA58) mutant of the SV40 large T antigen. Mol Cell Biochem 136(1):29–34
114. Eulalio A, Mano M, Dal Ferro M et al (2012) Functional screening identifies miRNAs inducing cardiac regeneration. Nature 492(7429):376–381. doi:10.1038/nature11739
115. Tian Y, Liu Y, Wang T et al (2015) A microRNA-hippo pathway that promotes cardiomyocyte proliferation and cardiac regeneration in mice. Sci Transl Med 7(279):279ra38. doi:10.1126/scitranslmed.3010841
116. Aguirre A, Montserrat N, Zachiggna S et al (2014) *In vivo* activation of a conserved microRNA program induces mammalian heart regeneration. Cell Stem Cell 15(5):589–604. doi:10.1016/j.stem.2014.10.003
117. Polizzotti BD, Ganapathy B, Walsh S et al (2015) Neuregulin stimulation of cardiomyocyte regeneration in mice and human myocardium reveals a therapeutic window. Sci Transl Med 7(281):281ra45. doi:10.1126/scitranslmed.aaa5171
118. Bersell K, Arab S, Haring B, Kühn B (2009) Neuregulin1/ErbB4 signaling induces cardiomyocyte proliferation and repair of heart injury. Cell 138(2):257–270. doi:10.1016/j.cell.2009.04.060
119. Dong J, Feldmann G, Huang J et al (2007) Elucidation of a universal size-control mechanism in Drosophila and mammals. Cell 130(6):1120–1133. doi:10.1016/j.cell.2007.07.019
120. Lin Z, von Gise A, Zhou P et al (2014) Cardiac-specific YAP activation improves cardiac function and survival in an experimental murine MI model. Circ Res 115(3):354–363. doi:10.1161/CIRCRESAHA.115.303632
121. Heallen T, Morikawa Y, Leach J et al (2013) Hippo signaling impedes adult heart regeneration. Development 140(23):4683–4690. doi:10.1242/dev.102798
122. Xin M, Kim Y, Sutherland LB et al (2013) Hippo pathway effector Yap promotes cardiac regeneration. Proc Natl Acad Sci U S A 110(34):13839–13844. doi:10.1073/pnas.1313192110
123. Mizuno T, Yau TM, Weisel RD, Kiani CG, Li R-K (2005) Elastin stabilizes an infarct and preserves ventricular function. Circulation 112(Suppl9):I81–I88. doi:10.1073/pnas.1313192110
124. Piek A, de Boer RA, Silljé HHW (2016) The fibrosis-cell death axis in heart failure. Heart Fail Rev 21(2):199–211. doi:10.1007/s10741-016-9536-9
125. Engler AJ, Carag-Krieger C, Johnson CP et al (2008) Embryonic cardiomyocytes beat best on a matrix with heart-like elasticity: scar-like rigidity inhibits beating. J Cell Sci 121(Pt 22):3794–3802. doi:10.1242/jcs.029678
126. Patwari P, Lee RT (2008) Mechanical control of tissue morphogenesis. Circ Res 103(3):234–343. doi:10.1161/CIRCRESAHA.108.175331
127. Hove JR, Köster RW, Forouhar AS et al (2003) Intracardiac fluid forces are an essential epigenetic factor for embryonic cardiogenesis. Nature 421(6919):172–177. doi:10.1038/nature01282
128. Jacot JG, Martin JC, Hunt DL (2010) Mechanobiology of cardiomyocyte development. J Biomech 43(1):93–98. doi:10.1016/j.jbiomech.2009.09.014
129. Arshi A, Nakashima Y, Nakano H et al (2013) Rigid microenvironments promote cardiac differentiation of mouse and human embryonic stem cells. Sci Technol Adv Mater 14(2):pii:025003. doi:10.1088/1468-6996/14/2/025003
130. Young JL, Kretchmer K, Ondeck MG, Zambon AC, Engler AJ (2014) Mechanosensitive kinases regulate stiffness-induced cardiomyocyte maturation. Sci Rep 14:6425. doi:10.1038/srep06425

131. Wei K, Serpooshan V, Hurtado C et al (2015) Epicardial FSTL1 reconstitution regenerates the adult mammalian heart. Nature 525(7570):479–485. doi:10.1038/nature15372
132. Reinecke H, Zhang M, Bartosek T, Murry CE (1999) Survival, integration, and differentiation of cardiomyocyte grafts: a study in normal and injured rat hearts. Circulation 100(2):193–202. doi:10.1161/01.CIR.100.2.193
133. Hong KU, Guo Y, Li Q-H, Cao P et al (2014) C-kit+ Cardiac stem cells alleviate post-myocardial infarction left ventricular dysfunction despite poor engraftment and negligible retention in the recipient heart. PLoS One 9(5), e96725. doi:10.1371/journal.pone.0096725
134. Santhakumar R, Vidyasekar P, Verma RS (2014) Cardiogel: a nano-matrix scaffold with potential application in cardiac regeneration using mesenchymal stem cells. PLoS One 9(12), e114697. doi:10.1371/journal.pone.0114697
135. Hasan A, Khattab A, Islam MA et al (2015) Injectable hydrogels for cardiac tissue repair after myocardial infarction. Adv Sci 2:1500122. doi:10.1002/advs.201500122
136. Chavakis E, Koyanagi M, Dimmeler S (2010) Enhancing the outcome of cell therapy for cardiac repair: progress from bench to bedside and back. Circulation 121(2):325–335. doi:10.1161/CIRCULATIONAHA.109.901405

Chapter 7
Regenerative Strategies for the Central Nervous System

Adam Roussas, Briana I. Martinez, Caroline P. Addington, and Sarah E. Stabenfeldt

7.1 Injury in the Central Nervous System: Physiology and Barriers to Regeneration

The central nervous system (brain and spinal cord; CNS) consists of complex molecular and cellular networks, which in turn renders the CNS sensitive to mechanical injury and neurodegenerative disease. Several CNS ailments, including traumatic brain injury (TBI), stroke, Alzheimer's disease, Parkinson's disease, and spinal cord injury (SCI) are increasing in prevalence in several countries, highlighting the need to delineate their pathophysiology [1–3]. In both the spine and brain, both degenerative and contusive injuries result in devastating, life changing results for patients such as partial or complete paralysis, neuropathic pain, and death [4]. SCI pathology is defined as the partial or complete paralysis and/or loss of sensation below the injury site, and TBI as perturbations in the brain that induce functional or cognitive disabilities. It is estimated that CNS injury impacts the lives of nearly six million people world-wide [5], with 276,000 of those individuals residing in the US [6]. Stroke and TBI patients also face great financial burden, paying hundreds of thousands of dollars per year, contributing to a total annual cost ranging from $30 to over $60 billion across the United States [7]. Further, patients expressing these pathologies report significantly lower quality of life, due in part to physical limitations and pain [8]. Although significant progress has been made in the pursuit of clinical methods to ameliorate injury progression, no viable technique to fully, or even partially, restore brain or spinal cord function after injury has emerged.

---

* Author contributed equally with all other contributors.

A. Roussas • B.I. Martinez • C.P. Addington • S.E. Stabenfeldt (✉)
School of Biological and Health Systems Engineering, Arizona State University, Tempe, AZ 85287, USA
e-mail: aroussas@asu.edu; bimarti1@asu.edu; caroline.addington@asu.edu; sarah.stabenfeldt@asu.edu

J. Wilson-Rawls, K. Kusumi (eds.), *Innovations in Molecular Mechanisms and Tissue Engineering*, Stem Cell Biology and Regenerative Medicine,
DOI 10.1007/978-3-319-44996-8_7

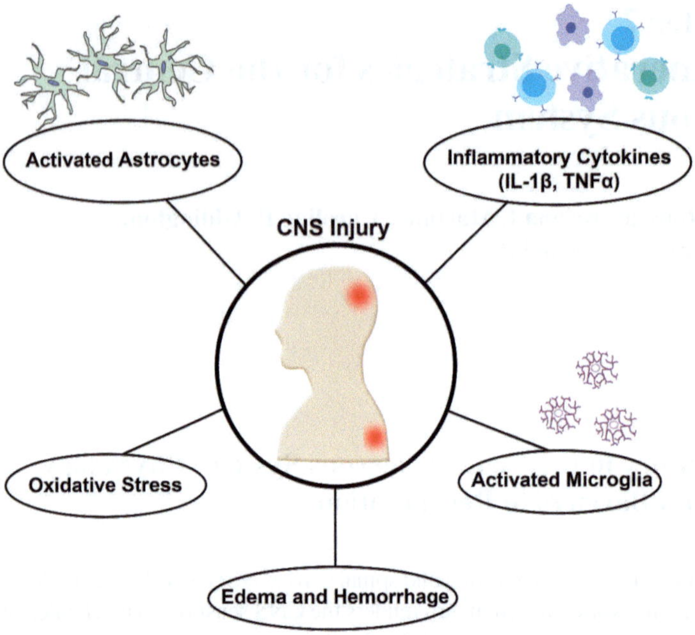

Fig. 7.1 General neuroinflammation cascade after CNS injury

Advances in neuroregenerative strategies are largely hindered by the complexity of the injury or disease pathology. For example, traumatic injury to the central nervous system, be it spinal cord or brain, stimulates a complex injury sequelae, commonly categorized into two major categories: the primary injury, known as the acute phase, and a more complex secondary injury (Fig. 7.1). Immediately following mechanical insult, the injury site swells with an influx of peripheral blood cells, various cytokines, and tissue debris that contribute to a hostile, neurotoxic environment [9]. Further, the swelling of soft tissue within a confined space (i.e. skull or vertebrae) leads to ischemia and cell death resulting in apoptosis of neurons and oligodendrocytes [9–11]. Together, these deleterious effects culminate in the progressive loss of neural function [9–12]. In the coming days to weeks, the inflammatory environment will continue to be stimulated and play an active role in shaping the secondary injury environment through loss of local vasculature and degeneration of surrounding myelinated axons and interneurons [9, 13]. Finally, via the interactions of a number of cytokines, growth factors, and astrocytes, a fluid-filled cyst lined with reactive astrocytes called the glial scar is left in place of the lesion. The scar acts as a barrier between the damaged area and healthy neural tissue, and the scar itself may extend beyond the lesion cavity boundary and acts as an impenetrable barrier for the growth of new axons [13, 14]. Many neuroscience and bioengineering research efforts have focused on developing methods to circumvent these barriers (i.e. exploring delivery options and modulating cellular environment). The purpose of this review is to discuss key approaches in neuroregeneration along with their benefits, limitations, and considerations for future research.

7.2 Manipulation of the Glial Scar

Local inflammation is an immediate consequence of neural injury and may lead to progressive cavitation and exacerbation of the primary lesion. As injury pathology progresses, local astrocytes activate to a reactive phenotype, exhibited by hypertrophy, in response to an intricate cascade of cytokine and growth factor signaling. The reactive astrocytes then form a dense scar tissue in an effort to protect intact neural networks from further damage [15, 16]. Although scar tissue is primarily produced by reactive astrocytes, the glial scar is a heterogeneous collection of many interacting cell types, forming a complex system of dystrophic axons, reactive astrocytes, stromal cells, activated microglia, and oligodendrocyte progenitors [17–23]. Leakage of blood and serum elements into the CNS parenchyma is considered an integral event in the formation of the glial scar. Most notably, however, astrocytes produce and deposit chondroitin sulfate proteoglycan (CSPG) throughout the extracellular matrix (ECM) within 24 h post-injury; high concentrations of CSPGs may persist at the injury site for months [24–27]. Although literature suggests that the glial scar acts to prevent propagation of the inflammatory response to healthy tissue [16, 28–30], it also serves as a significant barrier to axon regeneration [13, 14]. As such, many groups are seeking ways to break down or inhibit production and propagation of the glial scar. In particular, the direct administration (bolus or controlled release devices) of bacterial enzyme chondroitinase ABC (chABC) has shown to be effective in degrading the glial scar and promoting axonal growth by cleaving CSPGs in animal models of SCI [31–34].

7.2.1 *Direct Delivery of chABC*

Application of chABCs has been extensively studied in experimental brain and spinal cord injury models in attempts to delineate the enzyme's influence on the damaged CNS. chABC has traditionally been delivered to the injury site by bolus injection, as several groups have demonstrated the efficacy of intrathecal injections of chABC in murine models [31, 35–39]. Bradbury et al. first demonstrated that acute phase injections on alternating days up to 10 days post spinal cord injury was sufficient to promote CPSG degradation and functional recovery [31]. These results were later corroborated by both Barritt et al. and Cheng et al., where acute phase intrathecal injection of chABC post-SCI promoted CSPG degradation and subsequent axonal sprouting [37, 38]. Further, Cheng et al. demonstrated the dose dependence of chABC efficacy, with high dose groups exhibiting subarachnoid hemorrhages and death within 48 h of treatment [38]. In the brain, direct delivery via infusion after rodent models of TBI and nigrostriatal damage demonstrate the ability of chABC to locally degrade excessive CSPGs, thereby promoting axon regeneration [40, 41]. Moreover, bolus injection of chABC has demonstrated persistence in a rodent model of TBI, sustaining decreased inhibitory CSPG levels out to 28 days post injection [40]. Further, delayed chABC treatment in a rodent model of stroke demonstrated similar beneficial effects alongside behavioral recovery [31].

7.2.2 Controlled Release of chABC

In spite of promising results, diffusion of chABC into deep regions of the spinal cord and brain is limited when delivered intrathecally due to overflow beyond the intrathecal space and loss of bioactivity [42]. As such, controlled release systems (also see Sect. 7.6.1), such as hydrogel scaffolds and microspheres, have been explored as a method for direct, prolonged chABC administration. For example, Hyatt et al. demonstrated that controlled release from a fibrin delivery system afforded increased concentrations of bioactive chABC and enhanced CSPG degradation surrounding the lesion site compared to intraspinal injections [43]. Other groups have corroborated the use of natural hydrogel systems for sustained and controlled release of chABC to the injury site in murine models of SCI, recording stable release and augmented preservation of bioactivity [42, 44, 45]. Further, Huang et al. found that chABC loaded into poly-lactic acid microspheres is an effective method for preserving bioactivity and delivering chABC [45].

7.2.3 Genetic Engineering Approaches to Limit CSPG levels

Recently, researchers have explored genetic engineering approaches to imitate the effects chABC administration. Zhao et al. employed lentiviral vectors in a rat model of corticospinal tract lesion and observed significant degradation of CSPGs in the injured brain following intracortical injections of lentiviral vectors containing the chABC gene [46]. This reduction in CSPGs was correlated with marked reduction in axonal degeneration and augmented sprouting and short-range regeneration of corticospinal axons. The results of this study were corroborated by Bartus et al. [47]. An in vitro model of SCI using a Tet-On adenoviral vector encoding chABC also exhibited significant CSPG degradation in treatment groups [48]. In addition to increasing the expression of chABC, Donnelly et al. explored the effects of knocking down expression of one of the major pathways of CSPG formation, NG-2 [49]. They found that rats treated with short hairpin (sh) RNA designed to target NG-2 delivered by lentiviral vectors exhibited significantly reduced glial scar volume[49].

7.2.4 Drug Delivery Approaches to Enhance Neural Sprouting

Aside from high inhibitory levels of CSPG within the injury penumbra, additional inhibitory signals are readily present, including the neurite growth inhibitor Nogo-A, a myelin-associated protein. As such, researchers are investigating approaches to neutralize the inhibitory effects of Nogo-A. Infusion of Nogo-A antibodies 24 h after experimental stroke significantly increased axonal sprouting, subsequently promoting recovery from middle cerebral artery occlusion [50]. Moreover, delayed anti-Nogo-A treatment in a rodent model of stroke (7 days after infarct) significantly decreased infarct volume in comparison to animals with no treatment [51]. Ameliorating inflammation via Nogo-A modulation has also been observed in TBI. In a rat model of TBI, treatment with a Nogo-A antibody significantly increased axonal sprouting while decreasing behavioral deficits in comparison to a non-treatment group [52]. Several groups have assessed the viability of using Nogo-A antibodies to enhance neuroregeneration post injury in both murine and non-human primate models of SCI injury [53–56]. Caroni and Schwab first demonstrated the ability to augment neurite outgrowth and axonal growth using monoclonal antibodies to Nogo-A using cultured optic nerve explants [53], spurring many other groups to investigate the efficacy of Nogo-A antibodies. Leibscher et al. demonstrated the effectiveness of this technique in rat models of SCI, recording that antibody treated groups exhibited significantly enhanced regeneration of corticospinal neurons [54]. Soon after, the first non-human primate model of anti-Nogo-A administration was tested on Marmoset monkeys [55]. This group reported significantly enhanced sprouting and growth of lesioned spinal cord axons into and through the lesion site [55]. Freund et al. corroborated these findings in a Macaque monkey model [56]. A similarly promising approach for encouraging axonal growth is to target downstream intracellular signaling pathways such as the Rho/ROCK pathway, inhibition of which has been shown by a number of groups to promote axonal regeneration from neurites [57]. Rho/ROCK receptor antagonists like C3-exoenzyme, Y-27632, and ibuprofen have been associated with improved locomotor outcome in murine models of SCI. Several studies have found that administration of either Y-27632 or C3 transferase to inactivate Rho was sufficient to stimulate axon regeneration in SCI models [58–60]. Other groups corroborate the use of Y-27632 both in vivo and in vitro and also suggest that the effectiveness of treatment with Y-27632 is both dependent on dosage and timing: acute administration of high doses being most effective and low doses being potentially detrimental [60–63].

7.3 Modulation of the Inflammatory and Immune Response

As the inflammatory response progresses in the CNS after injury, the injury region is flooded with a myriad of inflammatory and immune response signaling factors. In an effort to induce neuroprotective signaling within the injury microenvironment and potentially mitigate the detrimental effects of CNS injury, molecular targets to modulate this response have been highly researched. For the purpose of this chapter, only a few specific factors will be discussed as they relate to modulating the injury microenvironment in both the spinal cord and brain: tumor necrosis factor alpha (TNFα), interleukin 1-β (IL-1β), vascular endothelial growth factor (VEGF), fibroblast growth factor 2 (FGF-2), and brain-derived neurotrophic factor (BDNF). The discussion will be centered on the role of various drugs/molecules, their delivery mechanisms, and how they modulate the inflammatory process after CNS injury.

7.3.1 *Administration of Cytokines, Growth Factors, and Neurotrophic Factors to Modulate Inflammation*

Cytokines are released into the inflammatory milieu by local activated glial cells and leukocytes and have been recorded to have both pro and anti-inflammatory effects in both the brain and spinal cord [64, 65]. Growth factors are naturally occurring proteins that promote cell proliferation, growth, and survival. Neurotrophins are a subset of these proteins and induce similar effects, specifically in neurons. Each of these proteins is a preferred ligand for a specific tyrosine kinase and therefore activates distinct signaling pathways [66, 67]. Cytokines, growth factors, and neurotrophic factors have been researched extensively in the injured or degenerating brain and spinal cord in attempts to provide new insights into the complex roles of these molecules in various neurological ailments. In this section, we discuss a select few of these molecules in terms of their functions and distinct avenues of their delivery. While we have selected only a small number of molecules that are relevant in both the brain and spinal cord, there are many other factors, such as chemokines and transcription factors, that significantly impact the inflammatory milieu, and it is important to note their effect on the injury environment as well.

7.3.2 *Tumor Necrosis Factor α*

Tumor necrosis factor α (TNFα) has primarily been characterized as a pro-inflammatory cytokine, inducing similar neurodegenerative and pro-inflammatory processes in both TBI and SCI [68–70]. Concentrations of TNFα have been found to peak at 1 h post injury in murine models of TBI/SCI, and as such, most research has focused on the acute phase of injury [64, 71–73]. At this early time point during the injury progression of both the brain and spinal cord, TNFα is expressed by all CNS cell types: microglia, astrocytes, neurons, and oligodendrocytes. While two weeks post injury, expression is primarily restricted to activated microglia and macrophages [74]. The major deleterious consequences of TNFα are apoptosis, ischemia, and glial cell activation. Acute inhibition of TNFα expression by either knockout or TNFα antagonist administration decreases edema, cortical tissue loss, and enhances performance on standard motor tasks after TBI and ischemic injury [70, 75, 76]. These data were corroborated in models using TNFα receptor (TNFR) knockout mice, where neuronal apoptosis was diminished in mice lacking TNFR after induced focal cerebral ischemia [76]. Similarly, TNFα was linked to apoptosis of both neurons and oligodendrocytes in murine contusion and crush SCI models [77, 78]. It is thought that TNFα acts to promote inflammation by inducing c-FOS in the spinal cord and nuclear factor k-light-chain-enhancer of activated B cells (NF-kB) in the brain, a protooncogene associated with apoptosis [79, 80] and a transcription factor linked to glial activation [81]. As a result of enhanced glial activation in the spinal cord, TNFα contributes to the initiation of Wallerian degeneration (the disruption of myelin sheaths) [72, 82]. TNFα induces this degeneration via the activation of microglia at the injury site, which will then begin to phagocytose myelin at an augmented rate [83]. Previous studies identified a direct correlation between TNFα levels and the rate of Wallerian degeneration [82, 83].

In spite of the neurodegenerative effects discussed above, some evidence suggests that TNFα does offer some level of neuroprotection as well. For example, Mattson et al. reported in vitro protection of cultured hippocampal and neocortical astrocytes by TNFα under glucose deprivation and glutamate toxicity. Moreover, Mattson et al. demonstrated upregulation of calbindin, a calcium binding protein, in TNFα-treated cells, which may have suppressed elevation of intracellular calcium and conferred resistance to the glutamate insult [84]. Other studies corroborate this information and suggest that the neuroprotective and pro-inflammatory effects of TNFα act in a temporally dependent manner. In a cortical contusion TBI model, Scherbel et al. reported a biphasic response to injury from TNFα deficient mice when compared to wild type controls. At 24–48 h following cortical contusion, the knockout mice recovered faster than the respective controls; however, between 1 and 4 weeks they demonstrated greater neurological dysfunction [70]. In contrast, Bruce et al. reported that when evaluated 24 h after middle cerebral artery occlusion (MCAO), infract area and oxidative stress in TNFR deficient mice were significantly higher than wild type controls [76]. This biphasic trend has also been demonstrated in spinal cord models of neural injury. Chi et al. observed that in the acute phase post-SCI, transgenic rats over-expressing TNFα exhibited significantly higher levels of apoptotic cells, while in the chronic phase, TNFα over-expressing rats displayed improved tissue healing and more activated astrocytes on the lesion border compared to wild type controls[85]. Taken together, these studies suggest that TNFα exerts a toxic effect in the acute stage of inflammation, while the absence of TNFα is deleterious in the chronic stage of inflammation [86].

7.3.3 *Interleukin-1β*

Interleukin-1β (IL-1β) is a pro-inflammatory cytokine whose expression is greatly enhanced after injury in both the brain and spinal cord. Similar to TNFα, IL-1β is expressed by astrocytes and microglia in the brain 3–8 h after injury occurs, and its presence sharply decreases after 1–2 days, as evidenced by controlled cortical impact (CCI) and moderate fluid percussion injury (FPI) models [64, 87, 88]. While the same cells express IL-1β in the spinal cord, its expression peaks at 12 h and then immediately begins to decrease thereafter [74]. The primary function of IL-1β is to promote astrogliosis and initiate an array of pro-inflammatory responses [89, 90] and/or promote angiogenesis, neurogenesis, and leukocyte infiltration [91, 92] within the injury environment. IL-1β also activates microglia and endothelial cells, which in turn potentiates IL-1β's action on all affected cells [92–94]. Studies in IL-1β receptor (IL-1R) knock out (KO) mice have provided evidence of the events discussed above. In the brain, Basu et al. found that the presence of microglia/

macrophages and astrocytes after penetrating brain injury was significantly reduced in IL-1R KO mice as compared to wild type controls[91]. These findings were also associated with depressed basal levels of IL-1β itself [91]. A similar study looking at the spinal cord observed decreased macrophage recruitment and TNFα expression after spinal cord transection in IL-1R KO mice compared to wild type controls[95]. Intracerebroventricular injection of IL-1β was also strongly implicated in inducing neutrophil infiltration as compared to saline injected controls and IL-1R KO mice [93]. This finding was further corroborated by a study that induced chronic infection with *Trypanosoma brucei* and observed limited leukocyte infiltration in IL-1R KO mice [93]. Due to its effects on local inflammatory cell types, IL-1β is also associated with Wallerian degeneration in the spinal cord. Perrin et al. found that microinjection of IL-1β into the dorsal column white matter 5 days after dorsal hemisection resulted in significantly increased recruitment of microglia/macrophages and rapid clearance of myelin 9 days later [96].

IL-1β also acts indirectly on local cell types via modulating the expression of growth factors like BDNF [95, 97, 98] and FGF [99] and has been shown to increase lesion size after both TBI and SCI in correlation with increases in these growth factors [93–95, 97–101]. A study of rat hippocampal formations revealed that direct administration of IL-1β or lipopolysaccharide (LPS) (which potentiates IL-1β) was sufficient to decrease BDNF mRNA levels [97]. Tong et al. further validated these effects by demonstrating that IL-1β can interrupt the neuroprotective effects of BDNF by directly disrupting BDNF's signal transduction pathway [98, 101]. Conversely, an in vitro investigation found that direct administration of IL-1β augmented production of FGF by astrocytes and microglia [99].

7.3.4 Vascular Endothelial Growth Factor

Vascular endothelial growth factor (VEGF) is a secreted mitogen that significantly impacts the development of vascular networks and other endogenous repair mechanisms. For example, administration of a VEGF antagonist post-ischemia/reperfusion injury in the mouse brain significantly decreased infarct size compared to a control group [102]. Additionally, using VEGF-A knockout mice, Argaw et al. observed a significant decrease in $CD45^+$ inflammatory cells compared to wild type controls in a multiple sclerosis model [103]. In general, administration of VEGF post SCI or TBI reportedly promotes anti-inflammatory effects in the injury environment such as inhibition of inflammatory cytokines, amelioration of the cytotoxic injury environment, and induction of autophagy. Specifically, VEGF administration following LPS stimulated bacteria exposure to spinal neuroglia decreased protein expression of inflammatory cytokines IL-1β and TNFα [104]. The same study also linked VEGF administration with increased levels of autophagy proteins Beclin1 and LC3B, suggesting that VEGF administration may stabilize the injury microenvironment by inducing autophagy in local glia [104]. VEGF has also been associated with reducing the detrimental effects of glutamate-induced excitotoxicity and hypoxia/hypoglycemia in spinal cord neurons [105–107]. Although current studies were performed on motor neurons, these results may be beneficial for advancing knowledge of VEGF-mediated inflammatory modulation in the CNS as the pathways utilized in each study have direct connections to CNS injury, such as the ERK pathway, MAPK pathway, and phosphatidylinositol 3-kinase (PI3K) pathway.

7.3.5 Basic Fibroblast Growth Factor-2

Fibroblast growth factor-2 (FGF-2) has also been found to play an important role in decreasing inflammation and gliosis, amongst many other positive benefits via the ERK and PI3K pathways in the brain and spinal cord, respectively [108–113]. Ruffini et al. and Rottlaender et al. both report diminished inflammation via the reduction of multiple inflammatory cell types such as macrophages, microglia, and CD8-positive T-cells in murine encephalomyelitis models [114, 115]. Another study corroborated these findings in vitro, reporting that FGF-2 administration results in limited leukocyte migration [116]. Additionally, FGF-2 expression was significantly increased after gold ion injection in a cryo-lesion model of TBI, causing a significant decrease in activated microglia as well as an increase in cell proliferation in the subventricular zone [117]. There is also evidence linking FGF-2 to modulation of astrocytosis and gliosis, yet results have been contradictory in this respect. For example, in vitro studies have found that administration of FGF-2 significantly increases astrocyte migration and proliferation [118]. Goddard et al. reported that intraventricular injection of FGF-2 induced reactive gliosis, while Kasai et al. and Reilley at al. both demonstrated the inhibition of reactive gliosis with in vitro and in vivo models of SCI, using an intraventricular osmotic pump to provide growth factor in vivo [96, 119, 120]. Differences in results could depend on relative concentrations, model, and/or delivery methods used. Although FGF-2 may be a potent inhibitor of reactive astrocytosis and leukocyte migration to the injury area, there are significant barriers to its clinical use in the CNS, as it does not cross the blood-brain barrier (BBB) or blood-spinal cord barrier (BSCB) [121, 122]. Thus, FGF-2 administration is limited to either intrathecal injection, direct administration to the lesion site, or potentially via biomaterial-based micro- or nanocarriers.

7.3.6 Brain-Derived Neurotrophic Factor

Brain-derived neurotrophic factor (BDNF) is a neurotrophic growth factor that plays a significant role in both the brain and spinal cord. With respect to neuroinflammation, BDNF has been shown to have both pro- and anti-inflammatory effects in the injured CNS. In the brain, the Jiang group observed upregulation of inflammatory cytokines IL-10 and TNFα after BDNF treatment in murine stroke models stroke as compared to injured animals without treatment [123]. In contrast, BDNF and FGF-2 administered in combination have been shown to ameliorate neuroinflammation in an experimental model of epilepsy by decreasing astrocytosis, microcytosis, and IL-1β levels [124]. Further, it has been shown that increased concentration of BDNF can be induced by high peroxide concentrations in the rodent model of focal cerebral ischemia, which, in turn, reduces peroxide levels at the injury site [125]. Although BDNF is primarily associated with modulating inflammation in the brain, there has been significant work demonstrating its relevance to spinal cord pathologies as well. In vivo studies have reported that BDNF downregulates nitric oxide synthase (NOS) in damaged neurons after spinal cord injury, leading to decreases in free radical production and a more stable injury microenvironment [121, 122]. BDNF has also been reported to limit the accumulation of lipid peroxidation byproducts in injured spinal cord by manipulating microglial function, serving to prevent further oxidative damage [126, 127]. Moreover, BDNF may reduce BSCB breakdown, edema formation, and neuronal injury in the traumatized spinal cord in vivo [126].

7.4 The Role of Drug Delivery to Modulate Inflammation

Many groups have also successfully used drug delivery to manipulate the inflammatory response via cellular modulation and inhibition of gliosis. For instance, one approach blocks key chemotactic receptors on inflammatory cells, thereby limiting their inherent chemotactic response towards the injury site. Specifically, the chemokine antagonist, vMIPII, displays a broad spectrum of receptor activities and has been shown to bind with high affinity to various classes of chemokine receptors on many different inflammatory cells such as XCR, CCR, CXCR, and CX_3CR [128, 129]. However, vMIPII binding is not associated with the normal, rapid mobilization of calcium from intracellular stores and, furthermore, blocks calcium mobilization induced by endogenous chemokines [130], likely due to the inhibition of extravasation by hematogenous cells [128, 131]. Using rat models of stab wound injury and spinal cord contusion injury, Ghirnikar et al. reported that continuous infusion of vMIPII decreased infiltration of neutrophils, macrophages, and microglia at the site of injury [131]. Further, vMIPII infusion resulted in substantial reductions in neuronal loss and gliosis with concomitant increased expression of *Bcl2* gene [128], an endogenous inhibitor of apoptosis [132–135].

Other pharmaceutical approaches focus primarily on modulating leukocyte infiltration and inflammatory cytokine production. Such drugs include Lipitor, Imatinib, Rolipram, Thalidomide, and Minocycline [136–139]. A sphingosine receptor modulator, FTY720, has also been shown to inhibit leukocyte recruitment to the injury site when administered daily after spinal cord contusion [140]; however, the mechanism behind this action is still under investigation. Lipitor attenuates BSCB dysfunction by suppressing isoprenoid-dependent RhoA activation and preventing matrix metalloproteinase-9 (MMP9) expression, which results in reduced infiltration of neutrophils/macrophages and reduced expression of the inflammatory mediators TNFα and IL-1β

[136]. In addition to leukocyte infiltration modulation, Lipitor may also reduce axonal degeneration, myelin degeneration, gliosis, and neuronal apoptosis as well as enhance tissue sparing after spinal cord contusion [136].

Imatinib reportedly modulates tyrosine kinase signaling cascades involved in local inflammation. Imatinib has been shown to mediate cytokine production in mast cells, macrophages, and effector T cells via inhibition of the protooncogene c-Kit, macrophage colony stimulating factor (MCSF), and the lymphocyte-specific protein tyrosine kinase (LCK) pathway, respectively [141]. Administration of Imatinib after a contusion SCI model improved BSCB integrity and functional outcomes, attenuated astrogliosis, decreased deposition of CSPGs, and increased tissue sparing [138]. Positive effects have also been observed with Imatinib administration after TBI. Imatinib is an efficient antagonist of platelet-derived growth factor receptor-α (PDFR), a receptor that plays a vital role in BBB permeability [142]. Treatment with Imatinib in a rodent model of TBI inhibited the PDFR pathway and consequently decreased BBB leakage, edema formation, and lesion size in the rodent model of TBI and subarachnoid hemorrhage [142, 143].

Similarly, Rolipram, a phosphodiesterase-4 specific inhibitor, is a potent suppressor of TNFα and IL-1β expression from LPS-stimulated macrophages [144, 145]. Rolipram's anti-inflammatory effects are induced via elevated intracellular cAMP levels [146]. Further, Rolipram administration in the rat ventrolateral funiculus was found to save oligodendrocytes in contusive spinal injury [147, 148]. Data collected in other murine models corroborate these findings and suggest that intravenous (IV) injections of Rolipram 1 h post-injury increase neuronal and oligodendral survival [139]. This group also investigated the effects of IV, subcutaneous, and oral Rolipram administration in the spinal cord, concluding that IV administration yields the most potent effects [139]. In the brain, administration of Rolipram has shown promise in treating ailments that arise from focal cerebral ischemia. Researchers have observed reduced expression of IL-1β and TNFα as well as improved sensorimotor function in rodent stroke models [149, 150]. Additionally, Rolipram has been found to increase survival of newborn neurons in the hippocampus after stroke, possibly by sustaining activation of the cAMP-responsive element binding protein pathway, which regulates neurogenesis under pathological conditions [150]. While Rolipram has been effective in stroke research, studies of Rolipram administration after TBI have demonstrated unfavorable results. Even though Rolipram decreases pro-inflammatory cytokines after injury, administration worsened injury outcome by significantly increasing hemorrhage and infarct size compared to vehicle-treated animals [151, 152]. These data suggest that while Rolipram administration is beneficial for some CNS injuries, further investigation is required to delineate how Rolipram may modulate the inflammatory response in certain pathophysiological contexts.

Another group found that the combination of Rolipram and Thalidomide acts as a potent inhibitor of TNFα and IL-1β expression, leading to significant tissue sparing [153]. Thalidomide alone has been reported to readily cross the BBB, reduce the release of TNFα from LPS-stimulated macrophages, and promote production of IL-10, an anti-inflammatory cytokine [154]. Administration of Thalidomide, in similar fashion to Rolipram, demonstrates favorable outcomes in experimental models of stroke due to its ability to suppress TNFα and IL-1β inflammatory cytokines. Not only have researchers observed decreased infarct volume and motor control deficits, but also significant decreases in oxidative damage to the brain [155, 156]. Due to its success in models of inflammation in the CNS, the mechanisms of Thalidomide treatment in other CNS injuries would be of interest.

The anti-microbial drug minocycline, when administered in the acute phase, can modulate the behavior of microglia via caspase 1 inhibition to reduce the inflammatory response and maintain a pro-regenerative milieu, leading to enhanced rehabilitative outcomes in SCI contusion models in mice [157]. A similar effect was recorded in murine contusion models of the brain, where minocycline attenuated microglial activation in one study [158] and IL-1β expression in another [159]. Given these results and others in both the spinal cord [160, 161] and brain [159, 162–166], there is currently a clinical trial recruiting participants to assess the safety and feasibility of clinical minocycline use after TBI [165]. While the scientific community has witnessed the failure of over 100 different neuroprotective drugs to enhance recovery in treatment of SCI and TBI [167], these five are still promising in their own merit and may also serve to elucidate new pathways for research.

7.5 Promoting Neuroprotection and Neuroregeneration through Administration of Growth Factors, Neurotrophic Factors or Small Molecules

In the previous sections, we presented the barriers to regeneration in the CNS such as gliosis, ischemia, and induced apoptosis, and key examples of employing growth factors, neurotrophic factors, and drugs to ameliorate these processes with the ultimate goal of promoting a more favorable microenvironment for neuroprotection and neuroregeneration. In this section, we discuss the direct links between growth/neurotrophic factors and neuroprotection/neuroregeneration (see Table 7.1).

7.5.1 Vascular Endothelial Growth Factor

Vascular endothelial growth factor (VEGF) is induced by hypoxia and ischemia and plays a role in enhancing angiogenesis and aproviding neuroprotection in the brain through the extracellular signal-regulated kinase (ERK) and endoplasmic reticulum (ER) stress pathways [168–172]. This group has shown that the actions of VEGF are dose dependent (demonstrating efficacy at about 2.5 ng/μL) and act most effectively within the first 3 h of transient MCAO [173]. VEGF efficacy may also be temporally dependent as evidenced in a rodent model of stroke [170]. Chu et al. observed VEGF IV administration 1 h post-insult to increase BBB leakage and lesion size, while

Table 7.1 Cytokines and neurotrophic/signaling factors associated with their respective effects on the CNS post-injury

Pro-inflammatory | Anti-inflammatory | Endogenous Repair | in the brain | in the SC

Molecule	Effect		in the brain	in the SC	Effect		in the brain	in the SC
TNF-α	Increased apoptosis [70,75,77,78,76]	Pro-inflammatory	✓	✓	Prevents apoptosis[84]	Anti-inflammatory	✓	
	Promoted Ischemia[70,76]	Pro-inflammatory	✓		Reduced oxidative stress[76]	Anti-inflammatory	✓	
	Increased glial activation[70,76]	Pro-inflammatory	✓		Glial cell activation[82]	Anti-inflammatory		✓
	Promoted Wallerian degeneration[82,83]	Pro-inflammatory		✓				
IL-1β	Increased glial activation [91,74]	Pro-inflammatory	✓	✓	Increased FGF concentration[94]	Anti-inflammatory	✓	
	Enhanced leukocyte infiltration[91,92,93,94,95]	Pro-inflammatory	✓	✓	↓ BDNF expression[90,92,93,174,179]	Anti-inflammatory	✓	✓
	Promoted Wallerian degeneration[94]	Pro-inflammatory		✓				
VEGF	↑ endogenous cell proliferation [104,176,177,178]	Endogenous Repair	✓	✓	Prevents apoptosis[105,106,107,180]	Anti-inflammatory		✓
	Promotes angiogenesis[174]	Endogenous Repair	✓	✓	Ameliorates ischemia[171,172,173]	Anti-inflammatory	✓	
	Reduces lesion size[104,176,177,178,179]	Endogenous Repair		✓	Decreases IL-1β and TNF-α expression[109]	Anti-inflammatory		✓
	Promotes tissue sparing[104,176,177,178,179]	Endogenous Repair		✓	Decreases edema[102,103]	Anti-inflammatory	✓	
	Decreases leukocyte infiltration[103]	Anti-inflammatory	✓		Induces local autophagy[104]	Anti-inflammatory		✓
FGF-2	Enhanced stem cell proliferation[211,213]	Endogenous Repair	✓		↓Leukocyte/microglia infiltration[108,114,115,116]	Anti-inflammatory	✓	✓
	Decreased myelination[176]	Endogenous Repair	✓		Reduces Glial Cell Activation[117]	Anti-inflammatory	✓	
					Inhibits Reactive Gliosis[98,119,120]	Anti-inflammatory		✓
BDNF	Increases IL-1β and TNF-α expression[123,124]	Pro-inflammatory	✓		↓ peroxide accumulation[125,126,127]	Anti-inflammatory	✓	✓
	Reduce edema formation[126]	Anti-inflammatory		✓	Reduces BSCB breakdown[126]	Anti-inflammatory		✓
	Promotes axonal growth[8,19,,193]	Endogenous Repair		✓				
NT-3	Enhances neuronal survival[184,185,186]	Endogenous Repair		✓	Increases axonal growth[187,188,193]	Anti-inflammatory		✓
	Augments remyelination[194]	Endogenous Repair		✓				

administration 48 h post-insult increased angiogenesis [174]. Interestingly, routes of administration may also influence the effect of VEGF on infarct size following stroke in rodents. Kaya et al. observed IV administration of VEGF to increase lesion size, whereas intracerebroventricular administration of VEGF decreased lesion size and BBB permeability in the same animal model [175].

As a likely result of the anti-inflammatory properties of VEGF, several groups have demonstrated that VEGF significantly reduces lesion size, promotes endothelial cell proliferation, and promotes tissue sparing in ex vivo, in vitro, and in vivo models of SCI [104, 176–178]. Further, Kim et al. suggest that VEGF delivered by neural progenitor stem cells (NPSCs) genetically modified to overexpress VEGF enhanced the proliferation of glial progenitor cells and promoted angiogenesis and tissue sparing in an ex vivo model of SCI [177]. The neuroprotective effects of VEGF were further described by De Laporte et al., who demonstrated that in vivo, biomaterial-facilitated VEGF delivery in a rat hemisection SCI model enhanced tissue sparing and angiogenesis [179], with these effects exaggerated in the presence of FGF-2 [179]. These VEGF-mediated molecular and cellular changes have been observed by some to correlate with improvements in motor function after various murine SCI models. Liu et al. found that induction of VEGF with an engineered transcription factor after clip compression SCI promoted revascularization, decreased apoptosis, and was associated with greater functional outcomes for animals expressing VEGF compared to wild type animals [180]. Similarly, Nori et al. observed enhanced motor recovery when NPSCs were implanted after murine contusion SCI models, and these results were directly linked to presence of VEGF in treated animals as compared to saline injected controls [181]. In general, VEGF appears to promote neuroregeneration in the CNS: supporting the regeneration of brain and spinal cord microvasculature, axonal growth in the spinal cord, and direct neurotrophic effects in both the brain and spinal cord.

7.5.2 Brain-Derived Neurotrophic Factor and Neurotrophin-3

Other molecules found to be effective in preventing apoptosis are BDNF and neurotrophin-3 (NT-3) [155, 182]. Similar to BDNF, NT-3 is a neurotrophin in the nerve growth factor (NGF) family that is diversely expressed in the CNS, with greater expression in the spinal cord than the brain after injury. In fact, levels of NT-3 mRNA have been shown to decrease after hippocampal fluid percussion injury [178, 183]. Nonetheless, experimental induction of NT-3 has been shown to stabilize calcium concentrations and reduce apoptosis due to excitotoxic insults in the brain [183]. In the spinal cord, NT-3 is associated with promoting survival of endogenous neurons. For example, NT-3 significantly enhanced the survival of anterior horn neurons in mouse compression SCI models and significantly improved cell survival and reduced cell atrophy in both in vitro and in vivo models of SCI [184–186]. Significant data has also been collected associating NT-3 with increased plasticity, axonal growth, and augmented myelination post-injury [187–192]. As such, researchers have primarily investigated the pro-growth effects that NT-3 has on axons in the regenerating spinal cord. Early studies suggested that acute, sustained delivery of NT-3 promoted the growth of axons post-cortical lesion injury [187, 188, 193]; however, this growth did not continue beyond the lesion site [187]. Taylor et al. found that this barrier can be overcome by creating an NT-3 gradient that leads out of the lesion site [188], but growth for long distances was not attainable simply using a neurotrophic signal. A recent study performed by Hou et al. corroborates the data found by Taylor et al., and further suggests that continuous expression of NT-3 is essential for sustaining the viability and continued growth of new axons post-spinal lesion [189]. Further, there is some evidence to suggest that NT-3 can promote re-myelination when expressed by transplanted NPSCs [194].

Similarly, multiple studies have shown that BDNF promotes sustained axonal growth and sprouting [195–198]. Blesch et al. found that transient BDNF delivery is sufficient to sustain regenerated axons in spinal cord injury sites [196]. Sasaki et al. corroborated this finding in a rat model of SCI, where transplanting human mesenchymal stem cells (MSCs) genetically modified to overexpress BDNF augmented neuroprotection and axonal sprouting [198]. BDNF is also correlated with increased axonal growth and sprouting from transplanted cells in dorsal column lesions [195]. Further, both Nakajima et al. and Koda et al. found that in rat contusion models of SCI, treatment with exogenous BDNF was directly correlated with the survival and rescue of endogenous neurons [182, 199] and oligodendrocytes [182, 200].

In the brain, BDNF has been associated with enhancement of mesencephalic dopaminergic neurons and modulation of cognitive processes [201, 202]. Several studies have also demonstrated the importance of BDNF in promoting NPSC differentiation in vitro and in vivo after brain injury [172, 203]. Despite its widespread benefits in the brain, BDNF is not capable of crossing the BBB, rendering it ineffective when administered intravenously [204, 205]. Recently, a technique was proposed to circumvent this barrier by conjugating BDNF to a molecular Trojan horse [204]. The technique was successful in facilitating movement across the BBB and resulted in decreased infarct volume in a rodent model of stroke [204]. Similarly, modified NPSCs overexpressing BDNF have demonstrated promise as a delivery option, increasing local BDNF mRNA expression in a CCI model of TBI [206].

Together, these data suggest that BDNF and NT-3 are essential molecules for promoting sprouting and sustained axonal regrowth of both endogenous and transplanted cells. As such, various researchers have administered the molecules in combination with promising results. Simultaneous viral-mediated transfection of NT-3 and BDNF showed modest improvement over either growth factor individually in both in vitro and in vivo murine models of SCI [207]. Further, intrathecal infusion of NT-3 and BDNF together for 8 weeks was sufficient to promote a robust regeneration of spinal cord neurons into a nerve graft [184]. Several other studies found that direct delivery of BDNF and NT-3 significantly augmented axonal regeneration, myelination, and growth into semipermeable guidance channels in vivo [208–210]. Nonetheless, these molecules are short-lived within the injury site, and methods of prolonging bioactivity must be implemented in order to consider clinical translation.

7.5.3 Basic Fibroblast Growth Factor-2

FGF2 has mitogenic effects on neural progenitor cells within the subventricular zone (SVZ) and neurotrophic effects on dopaminergic neurons in the postnatal and adult brain [211–213]. In the non-injured postnatal and adult mouse brain, intraventricular administration of FGF2 enhances proliferation of neural progenitor cells and increases oligodendrocyte precursor (OPC) generation in the SVZ [211]. FGF-2 has, therefore, been considered as a therapeutic treatment in studies of Parkinson's disease and other age-related neurodegenerative diseases. In a rodent model of Parkinson's disease (PD), researchers observed increased OPC generation, enhanced survival of dopaminergic neuronal cultures, and protection from toxicity [213]. In stroke models, delayed administration of FGF-2 has been shown to decrease infarct volume and increase functional recovery in rats [37, 38]. Despite its benefits, FGF2 administration poses in vivo limitations. For example, while FGF-2 induced axon myelination in periventricular white matter, it also resulted in significant loss of oligodendrocytes at later time points in both healthy and PD brain models [211, 214]. Further, a significant decrease of myelination in the caudal anterior medullary velum has been reported as an effect of FGF-2 delivery to rat pups [43]. These data suggest that FGF2 has a dual effect on neural environment, both providing neuroprotection to endogenous cells and potentially limiting remyelination and oligodendrocyte survival. Therefore, investigations into parameters that may modulate FGF2 efficacy could inform the design of more precise FGF2 delivery paradigms.

7.6 Biomaterials to Enhance Neuroregeneration

In order to circumvent the barriers of therapeutic administration (e.g. BBB/BSCB permeability) and minimize invasive therapeutic delivery paradigms, researchers have turned to engineered biomaterials constructed from synthetic or natural materials. These techniques benefit the CNS immensely by providing increasingly efficient avenues for delivery of therapeutics for brain and spinal injury/disease pathologies.

There are three major requirements that a biomaterial must meet in order to be suitable in this regard. First, the mechanical properties of the biomaterial must be robust enough to sustain local fixation (specifically in the spinal cord), yet compliant enough so as not to compress the local tissue [215–217]. Second, the biomaterial must be sufficiently biocompatible so as to integrate with the local environment (i.e. appropriate porosity, permeability, and surface nanotopography) [216, 218]. Third, the material must degrade at a suitable rate, similar to that of the ingrowth of support tissue and the extension of extending axonal processes [216, 219]. Many different types of materials have been used to develop scaffolds for neuroregeneration including natural materials like hyaluronic acid (HA), collagen, chitosan, agarose, alginate, and more; synthetic materials, such as nitrocellulose membranes, synthetic polymers, and biodegradable synthetic polymers; and biological grafts, such as fetal tissue (brain and spinal cord) and peripheral nerve implants [215, 219–222].

While the great number of biomaterials currently in use experimentally seems to be individually idiosyncratic, they can be generalized into three categories: hydrogels, nanofibers, and micro/nano particles [220].

7.6.1 Hydrogels for Neuroregeneration

Hydrogels are networks composed of polymeric chains that act to mimic the extracellular matrix (ECM). Depending on the origin of the material, hydrogels are further stratified into natural, synthetic, and/or composite materials groups. Natural hydrogel systems include HA, collagen, chitosan, agarose, alginate, elastin, fibrinogen, laminin, gelatin, and more [223]. Examples of synthetic hydrogels readily employed in biomedical applications for CNS injury include poly(ε-caprolactone) (PCL), polyethylene glycol (PEG), poly(hydroxyethyl methacrylate) (PHEMA) [224], polyvinyl alcohol (PVA), and poly[N-(2-hydroxypropyl)methacrylamide] (PHPMA) [225]. In general, the physiochemical and structural characteristics of a hydrogel scaffold greatly influence cellular response. The most important characteristics that should be considered for any hydrogel are the hydrophilic properties, stiffness/elasticity, ligand density, and fiber orientation [226, 227]. Charge is a material characteristic that has also been well characterized and found to have significant effects on neurite extension and cellular morphology. For example, the length of neuron extension is directly proportional to the magnitude of the net positive charge on a scaffold on a certain mathematical domain [228], the high extreme of this domain leading to growth inhibition [229, 230]. Other studies also suggest that positively charged hydrogels are capable of sustaining both primary nerve cells and the neural support cells that are critical for regeneration [231]. Further, cell behaviors are determined by the balance of cell-cell adhesion and cell-substrate adhesion, which are a function of the hydrophilic or hydrophobic properties of the substrate itself and any bioadhesive domains [232–236].

The elasticity of a substrate influences cellular and axonal infiltration and tends to have an ideal range depending on cell type and ligand. For example, PC12 neurites, a rat adrenal pheochromocytoma cell line that is induced by NGF into a neuronal phenotype, were found to exhibit branching and outgrowth on fibronectin-based substrates with a shear modulus between 10 Pa and 10 kPa [237]. While another study found that PC12 cells on PEG substrate exhibited enhanced adhesion and outgrowth with increasing Young's modulus (between 75 and 400 kPa) [238]. Regardless of cell type or ligand, the literature is split on the relationship between the elastic modulus and extent of axonal infiltration [237, 239]. Variation in data is likely due, in part, to variation in ligand density. Engler et al. showed that cellular proliferation was greatly enhanced on substrates with higher collagen densities than on controls without collagen [240]. Similarly, Thomas et al. found that spreading and motility of malignant astrocytes on two-dimensional (2D) polyacrylamide followed a normal distribution with respect to both stiffness and concentration of bound collagen [241]. Further, recent studies have highlighted the importance of dimensionality in culture maintenance, as 3D matrices [239, 242] have been shown to critically affect the metabolic activity, growth, and phenotype of neural cell types [243, 244]. Further, cells in 2D cultures must reorganize their integrin cell surface receptors and cytoskeleton to adapt to the planar presentation of receptor ligands, which causes distinct dynamic and spatial differences in the distribution of cell-cell and cell-matrix interactions [245–247]. In general, dimensionality plays a major role in neurite extension, retraction, branching, and maturation into axons and dendrites. Further, neurons cultured in 3D versus 2D environments display distinctly different morphologies, as 3D cultures give rise to neuritic geometries that are more morphologically reminiscent of those that occur in vivo [244].

Substrate charge, stiffness, and dimensionality are also important regulators of stem cell phenotypic fate. Environmental stiffness is such a potent controller of cell fate that MSCs will differentiate into neuronal, muscle, and bone cells as gel stiffness is increased [248, 249]. Interestingly, however, when MSCs migrate from soft to stiff substrates, some cells preserve neural markers, suggesting that not just stiffness, but variation in stiffness may also impact cell fate [250]. Saha et al. observed neural progenitor/stem cells (NPSCs) to preferentially differentiate into neurons on 2D substrates of soft to intermediate stiffness and into astrocytes on stiffer substrates [251]. Likewise, 3D hydrogel systems with mechanical moduli similar to that of the brain were found to induce neuronal differentiation of NPSCs [252]. In addition to stiffness and dimensionality, there is some evidence to suggest that substrate charge plays a role in cell fate. For example, mouse embryoid bodies cultured on negatively and neutrally charged hydrogel substrates were found to differentiate into all three germ layers and just mesoderm, respectively [236]. Further, Hynes et al. present data that may indicate that neuronal differentiation of NPSCs is due, in part, to the charge of PLL hydrogels [227].

Another promising methodology for hydrogel nanotechnology is self-assembling peptides (SAPs). Such self-assembling systems facilitate non-invasive delivery directly into an irregular shaped lesion. SAPs aggregate in situ via van der Waals forces, hydrogen bonds, and electrostatic forces to form a stable network with minimal secondary damage [253, 254]. Many studies have demonstrated that a wide variety of peptides and proteins can be utilized to produce very stable and well-ordered nanofiber structures with exceptional regularity [255–258]. Additionally, SAPs will collapse into non-toxic L-amino acids, which can potentially be used by local cells for growth and repair [256]. The diameter of self-assembled nanofibers ranges from 10 to 100 times smaller than typical electrospun fibers (discussed in the next section), which is of particular relevance to tissue engineering. This property suggests that SAPs can provide cells with a more realistic 3D microenvironment [259]. Self-assembling nanofiber scaffolds have also been observed as a possible treatment to induce axonal growth as well as prevent significant lesioning of the brain in experimental brain injury [260, 261]. As such, a number of in vitro and in vivo studies have been performed to investigate the efficacy of SAPs in the context of neuroregenerative medicine.

SAP scaffolds can be chemically designed to incorporate specific functional ligands, such as integrin-binding epitopes, to enhance endogenous repair mechanisms. Of particular interest are the laminin epitope, IKVAV; the ionic self-complementary RADA epitope; and modifications thereof. SAPs containing IKVAV sequences have been found to suppress astrocytic differentiation from NPSCs and to promote neurite outgrowth from cultured neurons [262]. In vivo injections of IKVAV functionalized SAPs into a spinal cord compression model reduced astrogliosis and cell death at the injury site, increased the neuronal differentiation of NPSCs, and enhanced neurite outgrowth [262]. In a separate study, IKVAVA SAPs were demonstrated to inhibit glial scar formation and promote axon elongation in a murine model of SCI [263]. SAPs were administered 24 h after dorsoventral compression, and significant reductions in astrogliosis were observed in IKVAVA treated groups as compared to non-bioactive molecule treated groups [263]. Tysseling et al. observed that IKVAV peptide amphiphile (PA) injection promoted plasticity in serotonergic fibers, axon growth, and reduced the glial scar in rat contusion and mouse compression models of spinal cord injury [263, 264]. These findings may be due to the extremely high density of the IKVAV epitope within the scaffold (almost 10^3 greater than laminin) and differences in IKVAV versus laminin signaling mechanisms, though these suggestions require continued investigation to fully elucidate [262, 264]. Another study investigating RADA16-I functionalized SAPs found that not only did RADA16-I groups support attachment and differentiation of NPSCs in vitro and in vivo, but also served to bridge the injured spinal cord of rats after in vivo transplantation [265].

7.6.2 Nanofibers for Neuroregeneration

Nanofibers are porous networked fiber structures with individual fiber diameter of less than 1 μm that mimic the architecture of the ECM. The high surface area to volume ratio and extraordinary mechanical strength make nanofibers excellent materials for neuroregeneration applications. Compared to traditional biomaterials, nanofibers have the advantages of topography and porosity that mimic the naturally occurring extracellular matrix. Additionally, they exhibit excellent biocompatibility with low immunogenicity and are endowed with properties that help to bridge the lesion gap in transection injuries. Therefore, nanofibers serve as effective delivery systems for cellular grafts and/or therapeutic drugs. The major processing techniques available to produce nanofibers are electrospinning [266], molecular self-assembly [267], drawing out [268], and catalytic synthesis [269]. Electrospinning and self-assembling nanofibers are the most studied techniques for developing scaffolds for neural tissue engineering.

One of the major benefits of electrospinning is the ability to control fiber alignment, which has been shown to significantly impact neurite outgrowth, cellular proliferation, and cell fate. Parallel-aligned nanofibers have yielded increased rates of NPSC differentiation and neurite outgrowth along the direction of fiber orientation as compared to randomly orientated fibers from both MSCs and human embryonic stem cell (hESC)-derived NPSCs [270–272]. Similar studies demonstrated oriented neurite outgrowth and glial migration from dorsal root ganglia explants on a collagen/poly-ε-caprolactone blend and on poly-L-lactate electrospun scaffolds [270, 273]. Topographic alignment also affects cell phenotype. Specifically, Mahairaki et al. observed a significant increase in neuronal differentiation and neurite outgrowth in mouse embryonic stem cells [274]. In addition to spatial orientation, other physical properties of nanofibers, like nanofiber diameter and patterns, can be controlled in the production process and have been shown to have significant impacts on cell fate [275–277]. For example, one study found that compared to microfiber polylactic acid (PLA) scaffolds, nanofiber PLA scaffolds significantly increased neuronal differentiation of NPSCs [278]. Further studies demonstrated that NPSCs will selectively differentiate into oligodendrocytes on ~300 nm fibers, while displaying neuronal phenotypes on ~750 nm fibers [279]. Moreover, patterning, such as grooved substrates, increased cell alignment and neuronal differentiation of rat hippocampal progenitor cells as compared to randomly oriented scaffolds [271]. Consistent with these findings, it was later shown that nanoscale ridge/groove pattern arrays can effectively induce differentiation of hESCs and NPSCs into neuronal phenotypes without the addition of any biochemical or biological agents [278, 280].

Electrospun polymer nanofibers have also been shown to be useful in drug delivery applications [281]. Many therapeutic compounds can be easily incorporated into the electrospun polymers via the electrospinning process, which unlike common encapsulation techniques (discussed in the next section) does not require a complex preparation. There are two attractive properties of the electrospinning technique with respect to drug/bioactive material loading: (1) the molecular structure and bioactivity of the incorporated drugs/bioactive molecules are well maintained due to the mild processing conditions and (2) the burst release of drugs in vitro is greatly reduced. The drug release profile can be tailored to be rapid, immediate, delayed, or modified dissolution by changing the polymer carrier used [282]. Release systems are designed via two electrospinning methods. The first method for encapsulating drugs is via electrospinning core-shell structures: two miscible or immiscible components can be spun into a composite fiber with a core layer encapsulated inside a shell [283]. Drugs or bioactive materials encapsulated using this technique show steady release characteristics, sustaining relatively constant release up to 140 h with tunable initial release profiles [284, 285]. One in vitro release study indicated that threads made from the core-shell fibers could suppress the initial burst release and provide a sustained drug release profile that would be useful for administering growth factor or other therapeutic drugs [284]. Another group used emulsion electrospinning to develop a core-shell structure from ultrafine fibers of bovine serum albumin and poly(DL)-lactide that allowed for control over burst release profiles of bioactive protein, extending release to 3 months [286].

The second method is to mix both the drug(s) and polymer(s) together and perform the electrospinning process as normal, such that the drugs are embedded within the entanglement of fibers themselves. Using this method, drugs can be easily located on the surface of the fibers, resulting in a burst effect in the initial stage of drug release [287, 288]. To control drug release profiles, properties such as fiber diameter and drug loading can be modulated to yield either longer or shorter periods of specific release profiles. For example, increasing fiber diameter results in longer periods of zero order release, and higher amounts of encapsulated drug will result in a more significant burst release profile [289]. Chew et al. successfully stabilized human β-nerve growth factor (NGF) in an electrospun copolymer of g-caprolactone and ethyl ethylene phosphate with BSA as a carrier protein [286]. They reported that bioactive NGF sustained release for 3 months from the matrix, using a PC12 neurite outgrowth assay to confirm bioactivity of NGF [286]. Based on these data and others [284–286, 290], these electrospun nanomaterials may be ideal for growth factor and therapeutic drug localization.

The degradation of nanofiber matrices and subsequent release profiles have also been investigated as a function of local pathological pH environments. This approach allows for development of stimuli-responsive nanofibers, which, in turn, broadens their clinical relevance. In vitro studies on the release profile of pH-responsive electrospun nanofibers demonstrated that total amount of drug release was accelerated due to both the pH-induced structural and morphological changes of the drug-fiber complex and to degradation of the matrix polymers themselves [291, 292]. Additionally, modulating concentration of acid-labile polymer segments allows for further control over the burst release profile. Yuan et al. designed and produced an acid-responsive ibuprofen-loaded PLA fibrous scaffold doped with sodium bicarbonate to ameliorate the inflammatory response and promote regeneration. They reported reductions in gene expression of IL-6 and TNFα and increased expression of VEGF in a muscle wound. Given the critical roles of IL-6, TNFα, and VEGF in CNS injury pathology, the development of a similar system to treat CNS injury would be of interest to the field of neuroregeneration [293]. Further, results showed that the ibuprofen-loaded PLA fibrous scaffold attenuated the inflammatory response more effectively than no-drug and non-acid responsive controls [293].

7.6.3 Microparticles and Nanoparticles for Neuroregeneration

Nanoparticles (NPs) and microparticles (MPs) are colloidal submicron to micron sized polymeric particles, often with a therapeutic agent of interest encapsulated within the polymeric matrix or adsorbed or conjugated onto the surface of the structure [294, 295]. Traditionally, both NPs and MPs have been synthesized using techniques like emulsification, electrospraying, and microfluidics [296]. Although some authors reserve the term "nanoparticle" for specific size cut offs, for the purpose of this review, the terms "nanoparticle" and "microparticle" refer to particles where the dimensions of the particle are measured in nanometers and micrometers, respectively.

There have been a number of studies validating the efficacy of drug or bioactive molecule delivery in both the brain and spinal cord using biodegradable NPs (e.g. poly(lactide-co-glycolide) (PLGA and PLAs) and liposomes. PLGA and PLA are polyesters, which undergo hydrolysis upon implantation into the body, forming biologically compatible and easily metabolized moieties [297, 298]. Drug entrapped in PLGA and PLA is released at a sustained rate via diffusion of the drug in the polymer matrix and by degradation of the matrix itself [297]. In these systems, the rate of degradation can be modulated by either changing block co-polymer composition or molecular weight, which alters the release of encapsulated agent from days to months [299]. Already, this technology has been employed for innovation upon and advancement of current clinical technologies, such as the delivery of methylprednisolone.

Methylprednisolone (MPS) is used clinically to render neuroprotection by suppressing primary inflammation and lipid peroxidation when administered at high doses in the acute phase of SCI ranging from moderate to severe [300]. However, the use of MPS is controversial as there is evidence that systemic administration of high doses of MPS may cause pneumonia, sepsis, and death [301]. MPS-loaded NPs have been studied extensively to improve drug efficacy while neutralizing some of the detrimental side effects associated with systemic high doses. In hemi-section SCI models, both PLGA-NPs and carboxymethylchitosan/polyamidoamine dendrimers loaded with MPS demonstrate significantly improved outcomes including reduction in lesion size, suppression of microglial and astrocytic responses, and improved axon regeneration [302, 303]. It is likely that the low-dose (approximately 20x less than clinically relevant systemic doses) and reduction of freely circulating bioactive MPS are responsible for these improved capabilities and may potentially augment the safety of clinical MPS use. Others have investigated loading minocycline into polymeric polycaprolactone NPs [304]. Administration of minocycline-loaded NPs reduced the proliferation of microglia/macrophages and modulated their morphology from activated to resting in vitro [304]. Similarly, Racke et al. performed an in vitro comparison of treatment with minocycline-loaded PEGylated liposomes with daily minocycline injections for the treatment of CNS autoimmune diseases, finding that infrequent injections of PEG-minocylcline liposomes are an effective alternative pharmacotherapy to daily injections [305]. In addition to drugs, neurotrophic factors have been loaded into NPs and delivered to spinal cord lesion sites. YC Wang et al. performed intraspinal injections of glial cell-derived neurotrophic factor (GDNF) loaded PLGA-NPs in a rat contusion SCI model [306]. The group reported increased neuronal survival as a result of successful release of drug into the lesion site [306].

Given the innate drug delivery capabilities of NPs, particularly sustained intracellular retention, it has been postulated that NPs may provide practical vehicles for sustained gene transfer. While there has been a significant amount of research performed in other systems of the body, there is currently little work regarding NPs as gene transfer vehicles in the CNS. Lu et al. performed a study investigating liposome-mediated GDNF gene transfer to augment corticospinal tract recovery after SCI lesion [307]. The group found that in vivo transfer of GDNF cDNA promoted axonal regeneration and enhanced functional recovery, suggesting that liposomal-mediated delivery of cDNA may be a practical gene transfer method. The therapeutic efficacy imparted from NPs in this application is likely due to their ability to buffer therapeutic agents from degradation by lysosomal enzymes [308]. Hedley et al. demonstrated that DNA encapsulated in PLGA microspheres were protected from nuclease activity in vitro compared to non-encapsulated DNA [308].

Due to the larger diameter of microspheres, they are more generally used for cellular scaffolding and drug and bioactive factor delivery, with many of the same properties as NPs. Through the use of biodegradable polymeric MPs and various synthesis techniques, delivery characteristics can be finely tuned, where: (1) delivery increases with protein loading, (2) polymer degradation varies inversely with microsphere size, and (3) release profiles can be controlled by pore size [309–311]. Demonstrating the effect of protein loading on delivery time, Cao et al. delivered NGF from PLGA, PCL, and blended PCL/PLGA microspheres and varied the amount of protein loaded, with PCL encapsulating the greatest amount of protein and PLGA encapsulating the least [312]. They report that at the longest, bioactive NGF was detectable at 91 days in the PCL group, demonstrating how polymer characteristics can modulate delivery efficacy [312]. Benoit et al. confirmed these results in in vitro studies of NGF release from PLA and PLGA microspheres [313]. Burdick et al. produced similar results with CNTF, BDNF, and NT-3 loaded into PLGA microspheres, reporting neurotrophin burst release for the first 1–2 days followed by up to 3 weeks of near-linear release [314]. Further, microspheres loaded with neurotrophic or growth factors are commonly used to modulate stem cell behavior in experimental settings. For example, Kim et al. investigated both the in vitro and in vivo effects of dibutyryl cyclic-AMP (dbcAMP)-loaded PLGA microspheres on exogenous neural progenitor/stem cells (NPSCs) in a murine full transection model of SCI [315]. The authors recorded significantly improved NPSC survival in vivo and differentiation into neuronal lineages in dbcAMP-MP treated groups versus untreated groups [315]. In an interesting experiment by Ashton et al., PLGA microspheres were loaded with alginate lysase, and then administered to alginate hydrogels culturing NPSCs [316]. Alginate hydrogels are commonly used as 3D scaffolds for cell culture and transplantation, but can take months to resolve within implantation sites as mammals do not produce endogenous alginases [317, 318]. By loading PLGA MPs with alginate lysase, the authors demonstrate a controllable and tunable method for inducing enzymatic degradation of alginate hydrogels in vivo [316]. Further, the authors reported significantly augmented rates of NPSC expansion in PLGA MP groups as compared to non-degrading alginate hydrogels [316]. A more recent study demonstrated the ability of growth factor-loaded microspheres to mediate cellular behavior. Nie et al. demonstrated that transforming growth factor-beta1 (TGF-1β) loaded into PLGA microspheres promoted chondrocyte adhesion and growth on hydrogel scaffolds [311]. The authors also demonstrated tunable release profiles based on PLGA-MP pore size [311].

Microspheres themselves have also been used as scaffolds for cellular transplantation, as encapsulation of cells provides a protective barrier against host immune cell interactions after grafting. In two separate murine SCI model studies, Tobias et al. demonstrated that alginate-encapsulated BDNF-producing fibroblasts survived for 1 month in culture, produced bioactive neurotrophins, survived transplantation into the spinal cord of immunocompetent animals, and provided a permissive environment for local host axon growth [319]. These data build a case for use of microspheres as in vivo drug delivery vehicles, cellular grafting materials, and modulators of endogenous repair after brain and spinal cord injury.

7.7 Cellular Approaches

There is considerable data regarding the effect of exogenous cellular transplantation in CNS injury. Cell therapies may be delivered to the spinal cord and brain by direct injection, intrathecal infusion, polymeric microspheres, or biomaterial scaffolds [320, 321]. Cell therapies primarily aim to replace damaged endogenous cells, enhance the regeneration of endogenous tissues, and/or act as vehicles for gene delivery and growth and neurotrophic factor delivery [319, 322, 323]. Neural progenitor/stem cells (NPSCs) and mesenchymal stem cells (MSCs), are frequently utilized based on their multipotent nature and capacity to replace neuronal lineage cells, enhance axonal regeneration, and restore interneuron communication [324]. Somatic cells and tissues such as olfactory ensheathing cells (OECs), Schwann cells, fetal tissues, and peripheral nerves have been shown to be effective in decreasing excitotoxicity via the secretion of various growth and neurotrophic factors, producing a more favorable microenvironment for neuroregeneration [324]. While cellular therapies show great promise, there are a number of disadvantages to consider such as ethical issues, tumorigenicity, and immunological rejection [325]. These concerns vary with cell type, and as such, in this section we will discuss the advantages, disadvantages, and relevant applications of each cell type mentioned above in detail (see Table 7.2).

Table 7.2 Cell types associated with their respective effects on the CNS post-injury

Pro-inflammatory | Anti-inflammatory | Endogenous Repair | in the brain | in the SC

Molecule	Effect	
NPSCs	Increase axon ensheathing[334,335]	Prevents apoptosis[340,341]
	Promote axonal regrowth[334]	Decrease astrogliosis[343]
	Increase # endogenous oligodendrocytes[334,337]	Enhance motor/cognitive function[340,341,342]
	Secrete GDNF, NFG, BDNF[338,339]	Differentiate to neuronal lineages[338,339]
OPCs	Promotes remyelination[353,357,358,359,362]	Increased NSC proliferation[351]
	Prevents apoptosis[351]	
Schwann Cells	Promotes remyelination[366,367,368,369]	Promote axonal extension[366,367,368,369]
	Promotes tissue sparing[370]	Secretes NGF, FGF-2, BDNF, NT-3[370]
MSCs	Differentiate to neuronal lineage[380,378,379,380,384]	Secretes BDNF, FGF, NGF, Bcl-2[385,384,386,387]
	Increase endogenous cell proliferation[385,383]	↑ BDNF, FGF, NGF expression [370,383]
OECs	Promote axonal elongation[388,389,392]	
ADSCs	Simulates neurogenesis[395,400,402]	Promotes axonal elongation[400]
	Stimulates angiogenesis[400,402]	Secretes VEGF and BDNF[400,402,403]
	Enhances motor function recovery[395,405]	

7.7.1 Neural Progenitor/Stem Cells

NPSCs are multipotent, self-renewing precursor cells that give rise to astrocytes, oligodendrocytes, and neurons in the CNS. In the spinal cord, NPSCs are generally derived from fetal tissue. However, stem cells derived from fetal tissue (embryonic stem cells; ESCs) raise ethical concerns and are prone to tumor development experimentally [326]. In light of this, NPSCs have recently been discovered in the filum terminale of adult spinal cord tissue [327–330]. NPSCs derived from the adult spinal cord consistently produce a neuron to glia ratio of 3:1 [331, 332], and thus provide an attractive alternative to ESCs as a cell source capable of neuronal differentiation. NPSCs transplanted into the spinal cord after SCI have been reported to persist in the lesion site up to 24 weeks post injury [333] and have widely been shown to promote endogenous recovery after SCI lesion [334–336]. For example, Parr et al. demonstrated increased axon ensheathing by transplanted NPSCs at the injury site, increased numbers of endogenous oligodendrocytes, and notable axonal regrowth [334]. These results were further corroborated in non-human primate contusion SCI models performed by Nemati et al., who found that SVZ-derived NPSCs transplanted into the spinal cord of macaque monkeys selectively differentiated into neuronal lineages, homed to the injury site, and promoted improved behavioral outcomes [337]. NPSCs also protect against excitotoxicity and secrete neurotrophic factors such as GDNF, NGF, and BDNF [338, 339]. For example, Llado et al. observed in vitro data suggesting that murine NPSCs implanted adjacent to spinal cord organotypic sections will induce transplant-directed axonal outgrowth due to secretion of GDNF and NGF by transplanted cells [338]. Further, spinal cord explants were protected in the presence of NPSCs against glutamate induced neurotoxicity [339]. Similarly, Lu et al. found that NPSCs expressed detectable levels of GDNF, NGF, and BDNF both in vitro and in vivo, which facilitated host axonal growth when transplanted in a murine SCI model [338].

The presence of NPSC transplants in the striatum after brain injury is of particular interest, as it has been associated with enhanced motor and proprioceptive recovery in neurodegenerative diseases. For instance, both the Anderson and the Ebert groups employed amphetamine-induced rotation tasks to demonstrate that NPSC transplantation is associated with increased recovery of motor symmetry [340, 341]. Further, Shear et al. demonstrated that the presence of exogenous NPSCs in the injured hippocampus persisted up to 12 months post injury in a rodent model of CCI, which was associated with long-term motor and cognitive recovery compared to vehicle treatment and non-treatment groups [342]. Other experimental TBI studies have demonstrated that transplanted NPSCs decrease astroglial activation and activated microglial accumulation post-injury in a model of mechanical hippocampal injury [343]. Finally, no significant differences in efficacy (as determined by histological analysis) have been observed between adult and embryonic NPSCs, prompting the necessity for more investigation into translational applications [321, 344].

Nonetheless, there are several problems that limit the use of NPSCs in clinical applications. Generally, NPSCs are less proliferative in culture than other stem cell lines, and as such may be more difficult to expand into large cultures as required for clinical work [345, 346]. Further, there is evidence to suggest that the differentiation potential of NPSCs decreases with time in culture [347]. Lastly, direct differentiation of NPSCs into neuronal lineages with high purity is difficult, although progress has been made in increasing phenotypic purity by using human NPSCs (hNPSCs) as opposed to murine [331, 335, 347, 348]. For example, Pfeifer et al. demonstrated pre-clinical efficacy of transplanted adult autologous NPSCs in murine models of cervical spinal cord lesion [335]. In this study, SVZ-derived NPSCs from a single, small biopsy were transplanted with autologous skin fibroblasts. The authors report that within 8 weeks post-biopsy, over 3 million NPSCs were generated from the single biopsy, and that the culture exhibited very similar differentiation profiles to that of syngeneic neuronal progenitor cell grafts [335]. Further, NPSCs within autologous fibroblast co-grafts remained viable up to 4 weeks post-transplantation and supplanted cystic lesion defects [335]. Since these findings, several clinical trials have been conducted in both the spinal cord and the brain.

Phase I/II trials have been conducted to assess the efficacy of transplanting hNPSCs after thoracic SCI [344]. The study, authorized by SwissMedic regulatory authority, enrolled 12 subjects 3–12 months prior to cell transplantation and administered fixed doses of 20 million cells directly into the thoracic spinal cord adjacent to the injury [344]. The trial reported significant sensory gains in a majority of the subjects, with two of the seven patients who were enrolled with complete motor and sensory injuries being converted to incomplete injuries after the onset of voluntary toe movement [344]. Further, there is currently a phase I trial at Emory University investigating the safety of surgically transplanting spinal cord derived NPSCs for the treatment of ALS [349]. Although this study is still on going, a completed phase I trial at Oregon Health Sciences University validated the safety of surgically transplanting hNPSCs into ventricular and bilateral subcortical sites to treat neuronal ceroid lipofuscinosis (Batten disease) [344, 350]. Subjects of this study received injections of 500 million or 1 billion cells and immunosuppression for the course of the study. Although some patients succumbed to the severity of their disease as evidenced by post-mortem examinations, four year follow-up of surviving patients showed no safety concerns [344, 350]. These findings were corroborated in another completed phase I open label study conducted at the University of California, San Francisco. In this study, four subjects with severe connatal Pelizaeus-Merzbacher myelin disorder were given injections of 300 million cells in the frontal white matter of each hemisphere and received immunosuppression for the first 9 months after injection [344, 350]. MRI assessments did not reveal signs of inflammation, gliosis, ischemia, or cystic changes and diffusion tensor imaging verified donor-tissue myelin development in the patients [344, 350]. However, continued testing in controlled studies will be required to demonstrate clinical efficacy of hNPSC transplantation.

7.7.2 Oligodendrocyte Progenitor Cells

Oligodendrocyte progenitor cells (OPCs) are a subtype of glial cells found in the CNS and are particularly prevalent in the hippocampus and neocortex [351, 352]. The primary function of OPCs is to maintain oligodendrocyte populations, the myelinating glia in the central nervous system [347]. Because of this, OPCs have been used extensively as a method to produce myelin building blocks for repair of injured white matter in rodent models of TBI and SCI [353–356]. Some groups have found that injection of OPCs in spinal cord transplant experiments increased remyelination [357, 358], with one group seeing these results only 7 days post-injury in thoracic contusion SCI models [359]. OPC and oligodendrocyte-conditioned media have been shown to increase axonal length and augment tissue sparing in vitro [360]. Lastly, OPCs have been observed to survive, migrate, and differentiate into adult oligodendrocytes after transplantation into a complete transection spinal cord injury model [361]. Similarly, in the brain, OPCs transplanted into dysmyelinated mouse brains differentiated into oligodendrocytes and significantly increased axonal myelination [353]. Similar effects were observed with induced pluripotent stem cell (iPSC)-derived OPCs transplanted into a hypomyelinated mouse brain [362]. OPCs transplanted in a rodent model of periventricular leukomalacia not only significantly increased myelination, but also increased proliferation of NPSCs and decreased neuronal cell loss [351]. Nonetheless, there is currently no sustainable source of OPCs, and OPC transplants are difficult to maintain at high phenotypic purity, thus limiting current clinical translation [358, 361].

7.7.3 Schwann Cells

Schwann cells are the myelinating cells of the peripheral nervous system that sustain peripheral axon regeneration. Nonetheless, there is strong evidence suggesting that Schwann cells can facilitate CNS axon regeneration as well. Schwann cells have been shown to promote remyelination of CNS axons, reduce lesion cavitation, and express various trophic factors when delivered via injection or biomaterial scaffold into SCI lesions [210, 363–365]. Further, Schwann cells are easily isolated from peripheral nerves and expanded in vitro [363]. Some groups have found that treatment of spinal cord transection with grafted Schwann cells was sufficient to allow damaged axons to extend into implanted grafts and become myelinated; however, the axons were unable to leave the grafts distally and re-innervate caudally located tissues [366–368]. In a contusion SCI model, transplanted Schwann cells significantly reduced cavitation at the injury site and promoted remyelination of endogenous axons growing into the graft [369]. Other studies have corroborated findings of reduced cavitation and further suggest that transplantation of Schwann cells may promote tissue sparing and form a bridge across the lesion site [370]. It has been postulated that the mechanism by which Schwann cells promote axonal regeneration is associated with the secretion of trophic factors like NGF, FGF-2, BDNF, or NT-3 by Schwann cells [371]. Schwann cells remain biologically active in the CNS for long periods of time, and have been shown to survive, integrate, and support axonal growth up to 5 weeks after transplantation in rat contusion SCI models [370]. In addition, more recent studies have reported cell survival for up to 6 weeks after transplantation in multichannel scaffolds [372]. Nonetheless, Schwann cells also exhibit limited migration from the graft site as a result of their inability to coexist with or migrate beyond astrocytes, resulting in axonal stalls at the graft-host interface [373]. As such, it is unlikely that Schwann cells alone will be sufficient to stimulate neuroregenerative effects in the CNS, but may play integral roles in combinatorial approaches to repair damage in the brain or spinal cord.

7.7.4 Mesenchymal Stem Cells

MSCs are adult stem cells obtained from bone marrow, blood, adipose, and dental tissues. MSCs are quickly and easily expanded in vitro, are easily isolated, can maintain their viability after cryopreservation at −80 °C, are able to self-renew, and have been reported to differentiate into essentially all non-hematopoietic lineages such as osteoblasts, adipocytes, chondrocytes, myoblasts, and early progenitors of neural cells [374, 375]. In fact, MSCs have been demonstrated to adopt neuronal phenotypes in in vitro studies and after in vivo transplantation in contusion SCI, stroke, TBI, and neurodegenerative disease models [376–380]. As a transplant option for neuroregeneration, MSCs are particularly useful due to a lack of antigens that trigger detrimental graft-versus-host responses [381]. Further, MSCs themselves secrete a number of anti-inflammatory, anti-apoptotic, and trophic signaling factors that support axonal growth, remyelination, and protection from cellular apoptosis [380, 382]. These many positive characteristics make MSCs unique candidates for autologous transplantation in the CNS in place of tumorigenic, ethically questionable ESCs [326]. Furthermore, MSC treatment in the mouse brain was associated with enhanced survival outcomes after observed increases in proliferation of endogenous neurons and oligodendrocytes after CCI and induced ischemia, results of which were correlated with the expression of BDNF, FGF, Bcl2, and NGF by MSCs [383–385].

Although, these neurotrophic and anti-inflammatory effects have primarily been observed in the brain, a recent study of MSC grafts after spinal cord compression injury found that MSC transplants secrete NGF and promote significant tissue sparing within the lesion area [386]. Further, results showed that grafted rats exhibit significantly greater revascularization than non-grafted rats [386, 387]. Taken together, these data illustrate a role for MSC transplants in promoting endogenous repair of host tissue. MSC transplants may also serve as a means of molecular delivery as several research groups have genetically engineered MSCs to deliver neurotrophic factors, receptor kinases, and HGF in an effort to promote graft survival and regeneration of host tissue [325, 386].

7.7.5 *Olfactory Ensheathing Cells*

Olfactory ensheathing cells (OECs) are glial cells found within both the peripheral and central nervous systems and are significant contributors to the regenerative capacity of olfactory neurons. In the CNS, OECs are found within the outer layers of the olfactory bulb, while in the PNS OECs are dispersed within the olfactory epithelium and the olfactory nerve. In either case, OECs form on bundles of olfactory sensory neuron axons, which are then able to extend and re-enter the olfactory bulb and re-synapse with second-order neurons in the glomerular layer [388]. The OECs interact with resident astrocytes and fibroblasts to facilitate these connections [388]. Further, OECs are capable of preventing axons from recognizing growth inhibitory molecules, thereby allowing them to elongate in otherwise inhibitory settings [389, 390]. As such, researchers have used these cells in various SCI models to promote axonal elongation. In a hemi-transection injury model, injected OECs induced axonal elongation into a denervated caudal host tract [391]. Similarly, in a transected adult rat SCI model, OECs significantly enhanced axonal regrowth, allowing axons to extend through white matter tracts, gray matter, and glial scars [390]. Other groups have found that using scaffolding techniques can augment OEC-mediated axonal growth. In one of the first studies utilizing this method, OECs induced axonal growth through a Schwann cell-containing channel for distances as long as 2.5 cm and to a slightly lesser extent in non Schwann cell seeded channels [392]. Given these promising results, there are now two successful clinical trials on record using OECs to promote functional recovery in spinal cord injury patients [393, 394].

7.7.6 *Adipose-Derived Stem Cells*

Adipose-derived stem cells (ADSCs) are more abundant, safer, and can be obtained in a relatively non-invasive manner compared to other common stem cells [395, 396]. Both in vivo and in vitro studies demonstrate that the presence of ADSCs supports neurogenesis and survival of neural stem cells, illustrating their neuroprotective benefits in addition to their safety [397–399]. Further, ADSCs appear to be safe for use in the spinal cord, as human ADSCs (hADSCs) transplanted into both humans and animals under various injury models showed no signs of tumorigenicity or adverse effects 3 months post-transplantation [396]. ADSCs have been recorded to differentiate into endothelial cells [400] in vitro after induced hypoxia, support axonal sprouting, and modify the structure of the glial scar in white matter after full transection [401]. Because ADSCs are so easily harvested and safe for human use, experiments have been performed via a number of neurological disorders to assess the therapeutic efficacy of these cells. For instance, research employing a murine model of Alzheimer's disease found that transplanted ADSCs both increase neurogenic activity in the SGZ and SVZ neurogenic niches and decrease the amount of oxidative stress on the neural environment [395]. Transplantation of these stem cells has also shown promise in treatment of stroke by reducing infarct size and promoting recovery of motor processes [395]. However, the mechanisms behind ADSC-mediated post-injury enhancement of motor function have yet to be fully elucidated in injury models of the brain.

In the spinal cord, there is evidence to suggest that partial recovery of motor function after SCI may be due to the stimulation of angiogenesis and neurogenesis by ADSCs [399, 402]. An in vitro study of ADSCs transplanted into simulated hypoxic conditions revealed enhanced neovascular formations, axonal growth, which correlated with significant functional recovery after in vivo transplantation into a rat SCI model [399]. These results were likely due to secretion of VEGF by ADSCs, as NPSCs co-transplanted with ADSCs exhibited significantly reduced apoptosis in the presence of ADSCs, while treatment with anti-VEGF attenuated this effect in a dose-dependent manner [399]. Zhou et al. corroborated these findings in rats that underwent bilateral dorsal laminectomy [402]. Human ADSCs (hADSCs) themselves exhibited elevated expression of VEGF and BDNF as compared to transplanted human bone marrow stem cells (hBMSCs) [402]. This expression was correlated with marked increases in angiogenesis and axon preservation and decreases in local activation of macrophages/astrocytes and lesion cavity formation in hADSC treated rats as compared to hBMSC rats [402]. Further, ADSC-derived Schwann cells have been shown to express a wide range of neurotrophic factors including NGF, BDNF, GDNF, and neurotrophin-4 (NT-4) [403]. Zainy et al. observed ADSC-derived Schwann cells modulate the hostile environment in a full transection SCI model to support axon regeneration and enhance functional recovery via the secretion of these molecules [404]. Finally, there are successful phase 1 and phase 2 clinical trials demonstrating the safety of autologous ADSC transplantation in acute spinal cord injury to improve functional outcome of treated patients [405].

7.8 Combinatorial Techniques to Enhance Neuroregeneration

Although there are many promising techniques to promote neural and endogenous regeneration after central nervous system injury, no specific technique has proved to be all encompassing in treating the number of factors impeding neuroregeneration in the CNS. As such, it is thought that a multifactorial approach, utilizing the desirable attributes of all the current methodologies and applying them in a simultaneous treatment, may be beneficial. As discussed, biomaterials can serve as excellent delivery vehicles for drugs, bioactive factors, and cells while providing physical support for grafted cells to ensure retention and distribution at the transplantation site. Matrices like these may enhance cell survival post-transplantation and promote differentiation into desired phenotypes based on the scaffolds properties. As such, many groups have attempted to incorporate combinations of neurotrophic signaling, drug delivery, cellular delivery, and hydrogel scaffolding in one treatment approach with varying results. Our review found that while modest attempts have been performed to combine various drugs, cells, and biomaterial implants in the brain, researchers have attempted many more combinatorial approaches in the spinal cord.

The administration of VEGF following transplantation of NPSCs in the rodent model of stroke, for example, has demonstrated significant motor recovery compared to groups that were only treated with VEGF or NPSCs alone [406]. Combinatorial treatment with NPSCs has also been extended to an experimental model of hypoxic-ischemia, where transplantation of progenitor cells combined with chABC significantly decreased infarction size compared to groups without combined therapy [407]. Additionally, growth factors have also been encapsulated in hydrogels in attempts to increase migration of neurons and have even been utilized to engineer specialized scaffolds to deliver stem cells [408, 409]. Nonetheless, there are limitations to consider when experimenting with the simultaneous treatment of several techniques. For example, VEGF co-delivered with FGF after closed head injury had no significant effects compared to groups that received a single growth factor alone [410]. The authors of this study hypothesize that signaling pathways may become oversaturated in response to elevated concentrations of various signaling factors, which could be potentially problematic for many combinatorial approaches [410].

As a proof of concept for the use of combinatorial treatments in the spinal cord, Johnson et al. investigated the efficacy of NPSCs transplanted in fibrin scaffolds impregnated with growth factors to enhance cell survival and promote neuronal differentiation [411]. The authors report that the combination of NT-3, PDGF and fibrin scaffold supported NPSC activity up to 8 weeks after transplantation and was successful in significantly increasing NPSC retention in vivo as compared to bolus and growth factor-free scaffold transplant conditions [411]. Later studies would then investigate similar combination treatment methods on the regenerative capabilities of the spinal cord in addition to cellular behavior, with varying results. For example, Kim et al. loaded PLGA microspheres with dbcAMP and separately cultured NPSCs on fibrin scaffolds, which were then both seeded onto chitosan microconduits to study the effects of NPSC behavior in vitro and in vivo [315]. Although it was found that transplanted NPSC/microsphere loaded microconduits were successful in promoting NPSC survival and neuronal differentiation, the data suggest that pretreatment with dbcAMP, but not microsphere treatment, increased in vivo survival and neuronal differentiation, suggesting that dbcAMP alone may be sufficient to induce these effects [315]. Nonetheless, the full combination strategy of stem cell and microsphere loaded chitosan channels was still effective in promoting NPSC survival and differentiation, and promoted extensive host axonal regeneration and improved recovery 6 weeks after full transection of the spinal cord [315]. In a similar study, Wilems et al. modified fibrin scaffolds with PLGA microspheres and encapsulated progenitor motor neurons (pMNs) in a model of rat sub-acute SCI [392]. PLGA microspheres were designed to sustain delivery of chABC and/or NEP1-40, a small myelin-associated inhibitor antagonist, for two weeks. While in vitro experiments confirmed that pMN viability was unaffected when cultured with chABC and/or NEP1-40, in vivo experiments with both molecules and encapsulated pMNs yielded reduced cell survival and increased macrophage infiltration. Further, scaffolds loaded only with pMNs decreased apoptosis and neuronal differentiation while simultaneously promoting axonal elongation and transplant integration into the host tissue [412]. While these studies do show promise for the effectiveness of combinatorial methods using only biomaterials, trophic factors, and a single cell type, more work is necessary to fine-tune the sensitive interactions between carrier, signaling factors, and cells.

Other groups have provided evidence that multiple neural cell types and neurotrophic factors transplanted in biomaterial scaffolds may support cellular transplantation. For example, co-transplantation of NPSCs and Schwann cells in electrospun PLGA scaffolds has been shown to enhance axonal regeneration in several studies [413–415]. Xia et al. demonstrated that NPSCs and Schwann cells in PLGA scaffolds promote axonal elongation in vivo; however, there was no differentiation of NPSCs into neuronal phenotypes in groups transplanted with Schwann cells, and axons were not able to form synaptic connections. Xiong et al. expanded on this approach by co-seeding NPSCs and Schwann cells in NT-3 loaded PLGA scaffolds in vitro [413]. The authors reported increased differentiation of NPSCSs into neurons and enhanced formation of active synaptic connections and myelination of neurites by the accompanied Schwann cells in vitro [413]. Taken together, this information suggests that while combinatorial techniques may indeed be capable of addressing current therapeutic limitations, there is still work to be done in determining the most effective therapeutic combinations.

References

1. Langlois JA, Rutland-Brown W, Thomas KE (2006) Traumatic brain injury in the United States: emergency department visits, hospitalizations, and deaths. Department of Health and Human Services, Centers for Disease Control and Prevention, Division of Acute Care, Rehabilitation Research and Disability Prevention, National Center for Injury Prevention and Control
2. Yarnall A, Archibald N, Burn D (2012) Parkinson's disease. Medicine (Baltimore) 40:529–535. doi:10.1016/j.mpmed.2012.07.008
3. Mergenthaler P, Dirnagl U, Meisel A (2004) Pathophysiology of stroke: lessons from animal models. Metab Brain Dis 19:151–167. doi:10.1023/B:MEBR.0000043966.46964.e6
4. Hagen EM, Lie SA, Rekand T et al (2010) Mortality after traumatic spinal cord injury: 50 years of follow-up. J Neurol Neurosurg Psychiatry 81:368–373
5. Tator CH (1998) Biology of neurological recovery and functional restoration after spinal cord injury. Neurosurgery 42:696–707
6. The National Spinal Cord Injury Statistical Center (2015) Spinal cord injury (SCI) facts and figures at a glance. Retrieved from: https://www.nscisc.uab.edu/Public/Facts%202015%20Aug.pdf
7. Ma VY, Chan L, Carruthers KJ (2014) Incidence, prevalence, costs, and impact on disability of common conditions requiring rehabilitation in the United States: stroke, spinal cord injury, traumatic brain injury, multiple sclerosis, osteoarthritis, rheumatoid arthritis, limb loss, and back pa. Arch Phys Med Rehabil 95:986.e1–995.e1. doi:10.1016/j.apmr.2013.10.032
8. Blesch A, Lu P, Tuszynski MH (2002) Neurotrophic factors, gene therapy, and neural stem cells for spinal cord repair. Brain Res Bull 57:833–838. doi:10.1016/S0361-9230(01)00774-2

9. Silva NA, Sousa N, Reis RL, Salgado AJ (2014) From basics to clinical: a comprehensive review on spinal cord injury. Prog Neurobiol 114:25–57. doi:10.1016/j.pneurobio.2013.11.002
10. Schwab ME (2002) Repairing the injured spinal cord. Science 295:1029–1031. doi:10.1126/science.1067840
11. Popovich PG, Wei P, Stokes BT (1997) Cellular inflammatory response after spinal cord injury in Sprague-Dawley and Lewis rats. J Comp Neurol 377:443–464
12. Yiu G, He Z (2006) Glial inhibition of CNS axon regeneration. Nat Rev Neurosci 7:617–627. doi:10.1038/nrn1956
13. Garbossa D, Boido M, Fontanella M et al (2012) Recent therapeutic strategies for spinal cord injury treatment: possible role of stem cells. Neurosurg Rev 35:293–311. doi:10.1007/s10143-012-0385-2, discussion 311
14. Cregg JM, DePaul MA, Filous AR et al (2014) Functional regeneration beyond the glial scar. Exp Neurol 253:197–207. doi:10.1016/j.expneurol.2013.12.024
15. Bush TG, Puvanachandra N, Horner CH et al (1999) Leukocyte infiltration, neuronal degeneration, and neurite outgrowth after ablation of scar-forming, reactive astrocytes in adult transgenic mice. Neuron 23:297–308. doi:10.1016/S0896-6273(00)80781-3
16. Faulkner JR, Herrmann JE, Woo MJ et al (2004) Reactive astrocytes protect tissue and preserve function after spinal cord injury. J Neurosci 24:2143–2155
17. Barnabé-Heider F, Göritz C, Sabelström H et al (2010) Origin of new glial cells in intact and injured adult spinal cord. Cell Stem Cell 7:470–482. doi:10.1016/j.stem.2010.07.014
18. Busch SA, Horn KP, Cuascut FX et al (2010) Adult NG2+ cells are permissive to neurite outgrowth and stabilize sensory axons during macrophage-induced axonal dieback after spinal cord injury. J Neurosci 30:255–265. doi:10.1523/JNEUROSCI.3705-09.2010
19. Göritz C, Dias DO, Tomilin N et al (2011) A pericyte origin of spinal cord scar tissue. Science 333:238–242. doi:10.1126/science.1203165
20. Meletis K, Barnabé-Heider F, Carlén M et al (2008) Spinal cord injury reveals multilineage differentiation of ependymal cells. PLoS Biol 6, e182. doi:10.1371/journal.pbio.0060182
21. Sabelström H, Stenudd M, Réu P et al (2013) Resident neural stem cells restrict tissue damage and neuronal loss after spinal cord injury in mice. Science 342:637–640. doi:10.1126/science.1242576
22. Soderblom C, Luo X, Blumenthal E et al (2013) Perivascular fibroblasts form the fibrotic scar after contusive spinal cord injury. J Neurosci 33:13882–13887. doi:10.1523/JNEUROSCI.2524-13.2013
23. Horn KP, Busch SA, Hawthorne AL et al (2008) Another barrier to regeneration in the CNS: activated macrophages induce extensive retraction of dystrophic axons through direct physical interactions. J Neurosci 28:9330–9341
24. Jones LL, Margolis RU, Tuszynski MH (2003) The chondroitin sulfate proteoglycans neurocan, brevican, phosphacan, and versican are differentially regulated following spinal cord injury. Exp Neurol 182:399–411. doi:10.1016/S0014-4886(03)00087-6
25. Tang X, Davies JE, Davies SJA (2003) Changes in distribution, cell associations, and protein expression levels of NG2, neurocan, phosphacan, brevican, versican V2, and tenascin-C during acute to chronic maturation of spinal cord scar tissue. J Neurosci Res 71:427–444. doi:10.1002/jnr.10523
26. McKeon RJ, Höke A, Silver J (1995) Injury-induced proteoglycans inhibit the potential for laminin-mediated axon growth on astrocytic scars. Exp Neurol 136:32–43. doi:10.1006/exnr.1995.1081
27. McKeon RJ, Jurynec MJ, Buck CR (1999) The chondroitin sulfate proteoglycans neurocan and phosphacan are expressed by reactive astrocytes in the chronic CNS glial scar. J Neurosci 19:10778–10788
28. Herrmann JE, Imura T, Song B et al (2008) STAT3 is a critical regulator of astrogliosis and scar formation after spinal cord injury. J Neurosci 28:7231–7243. doi:10.1523/JNEUROSCI.1709-08.2008

29. Okada S, Nakamura M, Katoh H et al (2006) Conditional ablation of Stat3 or Socs3 discloses a dual role for reactive astrocytes after spinal cord injury. Nat Med 12:829–834. doi:10.1038/nm1425
30. Wanner IB, Anderson MA, Song B et al (2013) Glial scar borders are formed by newly proliferated, elongated astrocytes that interact to corral inflammatory and fibrotic cells via STAT3-dependent mechanisms after spinal cord injury. J Neurosci 33:12870–12886
31. Bradbury EJ, Moon LDF, Popat RJ et al (2002) Chondroitinase ABC promotes functional recovery after spinal cord injury. Nature 416:636–640. doi:10.1038/416636a
32. Bradbury EJ, Carter LM (2011) Manipulating the glial scar: chondroitinase ABC as a therapy for spinal cord injury. Brain Res Bull 84:306–316. doi:10.1016/j.brainresbull.2010.06.015
33. Shields LBE, Zhang YP, Burke DA et al (2008) Benefit of chondroitinase ABC on sensory axon regeneration in a laceration model of spinal cord injury in the rat. Surg Neurol 69:568–577. doi:10.1016/j.surneu.2008.02.009, discussion 577
34. Cafferty WBJ, Yang S-H, Duffy PJ et al (2007) Functional axonal regeneration through astrocytic scar genetically modified to digest chondroitin sulfate proteoglycans. J Neurosci 27:2176–2185
35. Kuhn HG, Winkler J, Kempermann G et al (1997) Epidermal Growth Factor and Fibroblast Growth Factor-2 Have Different Effects on Neural Progenitors in the Adult Rat Brain. J Neurosci 17:5820–5829
36. Ninomiya M, Yamashita T, Araki N et al (2006) Enhanced neurogenesis in the ischemic striatum following EGF-induced expansion of transit-amplifying cells in the subventricular zone. Neurosci Lett 403:63–67. doi:10.1016/j.neulet.2006.04.039
37. Barritt AW, Davies M, Marchand F et al (2006) Chondroitinase ABC promotes sprouting of intact and injured spinal systems after spinal cord injury. J Neurosci 26:10856–10867. doi:10.1523/JNEUROSCI.2980-06.2006
38. Cheng C-H, Lin C-T, Lee M-J et al (2015) Local delivery of high-dose chondroitinase ABC in the sub-acute stage promotes axonal outgrowth and functional recovery after complete spinal cord transection. PLoS One 10, e0138705. doi:10.1371/journal.pone.0138705
39. Chen X, Liao S, Ye L et al (2014) Neuroprotective effect of chondroitinase ABC on primary and secondary brain injury after stroke in hypertensive rats. Brain Res 1543:324–333. doi:10.1016/j.brainres.2013.12.002
40. Lin R, Kwok JCF, Crespo D, Fawcett JW (2008) Chondroitinase ABC has a long-lasting effect on chondroitin sulphate glycosaminoglycan content in the injured rat brain. J Neurochem 104:400–408. doi:10.1111/j.1471-4159.2007.05066.x
41. Harris NG, Mironova YA, Hovda DA, Sutton RL (2010) Chondroitinase ABC enhances pericontusion axonal sprouting but does not confer robust improvements in behavioral recovery. J Neurotrauma 27:1971–1982. doi:10.1089/neu.2010.1470
42. Lee H, McKeon RJ, Bellamkonda RV (2010) Sustained delivery of thermostabilized chABC enhances axonal sprouting and functional recovery after spinal cord injury. Proc Natl Acad Sci U S A 107:3340–3345. doi:10.1073/pnas.0905437106
43. Hyatt AJT, Wang D, Kwok JC et al (2010) Controlled release of chondroitinase ABC from fibrin gel reduces the level of inhibitory glycosaminoglycan chains in lesioned spinal cord. J Control Release 147:24–29. doi:10.1016/j.jconrel.2010.06.026
44. Rossi F, Veglianese P, Santoro M et al (2012) Sustained delivery of chondroitinase ABC from hydrogel system. J Funct Biomater 3:199–208. doi:10.3390/jfb3010199
45. Huang Y-C, Hsu S-H, Chen M-T et al (2011) Controlled release of chondroitinase ABC in chitosan-based scaffolds and PDLLA microspheres. Carbohydr Polym 84:788–793. doi:10.1016/j.carbpol.2010.03.034
46. Zhao R-R, Muir EM, Alves JN et al (2011) Lentiviral vectors express chondroitinase ABC in cortical projections and promote sprouting of injured corticospinal axons. J Neurosci Methods 201:228–238. doi:10.1016/j.jneumeth.2011.08.003
47. Bartus K, James ND, Didangelos A et al (2014) Large-scale chondroitin sulfate proteoglycan digestion with chondroitinase gene therapy leads to reduced pathology and modulates mac-

rophage phenotype following spinal cord contusion injury. J Neurosci 34:4822–4836. doi:10.1523/JNEUROSCI.4369-13.2014

48. Curinga GM, Snow DM, Mashburn C et al (2007) Mammalian-produced chondroitinase AC mitigates axon inhibition by chondroitin sulfate proteoglycans. J Neurochem 102:275–288. doi:10.1111/j.1471-4159.2007.04530.x
49. Donnelly EM, Strappe PM, McGinley LM et al (2010) Lentiviral vector-mediated knockdown of the NG2 [corrected] proteoglycan or expression of neurotrophin-3 promotes neurite outgrowth in a cell culture model of the glial scar. J Gene Med 12:863–872. doi:10.1002/jgm.1509
50. Wiessner C, Bareyre FM, Allegrini PR et al (2003) Anti-Nogo-A antibody infusion 24 hours after experimental stroke improved behavioral outcome and corticospinal plasticity in normotensive and spontaneously hypertensive rats. J Cereb Blood Flow Metab 23:154–165. doi:10.1097/00004647-200302000-00003
51. Lee J-K, Kim J-E, Sivula M, Strittmatter SM (2004) Nogo receptor antagonism promotes stroke recovery by enhancing axonal plasticity. J Neurosci 24:6209–6217. doi:10.1523/JNEUROSCI.1643-04.2004
52. Lenzlinger PM, Shimizu S, Marklund N et al (2005) Delayed inhibition of Nogo-A does not alter injury-induced axonal sprouting but enhances recovery of cognitive function following experimental traumatic brain injury in rats. Neuroscience 134:1047–1056
53. Caroni P, Schwab ME (1988) Antibody against myelin associated inhibitor of neurite growth neutralizes nonpermissive substrate properties of CNS white matter. Neuron 1:85–96. doi:10.1016/0896-6273(88)90212-7
54. Liebscher T, Schnell L, Schnell D et al (2005) Nogo-A antibody improves regeneration and locomotion of spinal cord-injured rats. Ann Neurol 58:706–719. doi:10.1002/ana.20627
55. Fouad K, Klusman I, Schwab ME (2004) Regenerating corticospinal fibers in the Marmoset (*Callitrix jacchus*) after spinal cord lesion and treatment with the anti-Nogo-A antibody IN-1. Eur J Neurosci 20:2479–2482. doi:10.1111/j.1460-9568.2004.03716.x
56. Freund P, Schmidlin E, Wannier T et al (2006) Nogo-A-specific antibody treatment enhances sprouting and functional recovery after cervical lesion in adult primates. Nat Med 12:790–792. doi:10.1038/nm1436
57. McKerracher L, Higuchi H (2006) Targeting Rho to stimulate repair after spinal cord injury. J Neurotrauma 23:309–317. doi:10.1089/neu.2006.23.309
58. Dergham P, Ellezam B, Essagian C et al (2002) Rho signaling pathway targeted to promote spinal cord repair. J Neurosci 22:6570–6577
59. Monnier PP, Sierra A, Schwab JM et al (2003) The Rho/ROCK pathway mediates neurite growth-inhibitory activity associated with the chondroitin sulfate proteoglycans of the CNS glial scar. Mol Cell Neurosci 22:319–330. doi:10.1016/S1044-7431(02)00035-0
60. Sung J-K, Miao L, Calvert JW et al (2003) A possible role of RhoA/Rho-kinase in experimental spinal cord injury in rat. Brain Res 959:29–38. doi:10.1016/S0006-8993(02)03717-4
61. Chan CCM, Khodarahmi K, Liu J et al (2005) Dose-dependent beneficial and detrimental effects of ROCK inhibitor Y27632 on axonal sprouting and functional recovery after rat spinal cord injury. Exp Neurol 196:352–364. doi:10.1016/j.expneurol.2005.08.011
62. Borisoff JF, Chan CC, Hiebert GW et al (2003) Suppression of Rho-kinase activity promotes axonal growth on inhibitory CNS substrates. Mol Cell Neurosci 22:405–416. doi:10.1016/S1044-7431(02)00032-5
63. Nishio T (2009) Axonal regeneration and neural network reconstruction in mammalian CNS. J Neurol 256:306–309. doi:10.1007/s00415-009-5244-x
64. Addington C, Roussas A, Dutta D, Stabenfeldt S (2015) Endogenous Repair Signaling after Brain Injury and Complementary Bioengineering Approaches to Enhance Neural Regeneration. Biomark Insights 10:43. doi:10.4137/BMI.S20062.Received
65. Klusman I, Schwab ME (1997) Effects of pro-inflammatory cytokines in experimental spinal cord injury. Brain Res 762:173–184. doi:10.1016/S0006-8993(97)00381-8
66. Skaper SD (2012) The neurotrophin family of neurotrophic factors: an overview. Methods Mol Biol 846:1–12. doi:10.1007/978-1-61779-536-7_1

67. Patapoutian A, Reichardt LF (2001) Trk receptors: mediators of neurotrophin action. Curr Opin Neurobiol 11:272–280. doi:10.1016/S0959-4388(00)00208-7
68. Csuka E, Morganti-Kossmann MC, Lenzlinger PM et al (1999) IL-10 levels in cerebrospinal fluid and serum of patients with severe traumatic brain injury: relationship to IL-6, TNF-α, TGF-β1 and blood–brain barrier function. J Neuroimmunol 101:211–221. doi:10.1016/S0165-5728(99)00148-4
69. Lenzlinger PM, Morganti-Kossmann MC, Laurer HL, McIntosh TK (2001) The duality of the inflammatory response to traumatic brain injury. Mol Neurobiol 24:169–181. doi:10.1385/MN:24:1-3:169
70. Scherbel U, Raghupathi R, Nakamura M et al (1999) Differential acute and chronic responses of tumor necrosis factor-deficient mice to experimental brain injury. Proc Natl Acad Sci 96:8721–8726. doi:10.1073/pnas.96.15.8721
71. Taupin V, Toulmond S, Serrano A et al (1993) Increase in IL-6, IL-1 and TNF levels in rat brain following traumatic lesion. J Neuroimmunol 42:177–185. doi:10.1016/0165-5728(93)90008-M
72. Esposito E, Cuzzocrea S (2011) Anti-TNF therapy in the injured spinal cord. Trends Pharmacol Sci 32:107–115. doi:10.1016/j.tips.2010.11.009
73. Dinomais M, Stana L, Egon G et al (2009) Significant recovery of motor function in a patient with complete T7 paraplegia receiving etanercept. J Rehabil Med 41:286–288. doi:10.2340/16501977-0329
74. Pineau I, Sun L, Bastien D, Lacroix S (2010) Astrocytes initiate inflammation in the injured mouse spinal cord by promoting the entry of neutrophils and inflammatory monocytes in an IL-1 receptor/MyD88-dependent fashion. Brain Behav Immun 24:540–553. doi:10.1016/j.bbi.2009.11.007
75. Knoblach SM, Fan L, Faden AI (1999) Early neuronal expression of tumor necrosis factor-α after experimental brain injury contributes to neurological impairment. J Neuroimmunol 95:115–125. doi:10.1016/S0165-5728(98)00273-2
76. Bruce AJ, Boling W, Kindy MS et al (1996) Altered neuronal and microglial responses to excitotoxic and ischemic brain injury in mice lacking TNF receptors. Nat Med 2:788–794. doi:10.1038/nm0796-788
77. Yune TY, Chang MJ, Kim SJ et al (2003) Increased production of tumor necrosis factor-alpha induces apoptosis after traumatic spinal cord injury in rats. J Neurotrauma 20:207–219. doi:10.1089/08977150360547116
78. Lee YB, Yune TY, Baik SY et al (2000) Role of tumor necrosis factor-alpha in neuronal and glial apoptosis after spinal cord injury. Exp Neurol 166:190–195. doi:10.1006/exnr.2000.7494
79. Hermann GE, Rogers RC, Bresnahan JC, Beattie MS (2001) Tumor necrosis factor-alpha induces cFOS and strongly potentiates glutamate-mediated cell death in the rat spinal cord. Neurobiol Dis 8:590–599. doi:10.1006/nbdi.2001.0414
80. Zou JY, Crews FT (2005) TNF alpha potentiates glutamate neurotoxicity by inhibiting glutamate uptake in organotypic brain slice cultures: neuroprotection by NF kappa B inhibition. Brain Res 1034:11–24. doi:10.1016/j.brainres.2004.11.014
81. Khasnavis S, Jana A, Roy A et al (2012) Suppression of nuclear factor-kB activation and inflammation in microglia by physically modified saline. J Biol Chem 287:29529–29542. doi:10.1074/jbc.M111.338012
82. Sebastian J, Schroeter M (2003) Detrimental and beneficial effects of injury-induced inflammation and cytokine expression in the nervous system. Mol Cell Biol 513:87–113
83. Shamash S, Reichert F, Rotshenker S (2002) The cytokine network of wallerian degeneration: tumor necrosis factor-alpha, interleukin-1alpha, and interleukin-1beta. J Neurosci 22:3052–3060
84. Mattson MP, Cheng B, Baldwin SA et al (1995) Brain injury and tumor necrosis factors induce calbindin D-28k in astrocytes: evidence for a cytoprotective response. J Neurosci Res 42:357–370. doi:10.1002/jnr.490420310
85. Chi L-Y, Yu J, Zhu H et al (2008) The dual role of tumor necrosis factor-alpha in the pathophysiology of spinal cord injury. Neurosci Lett 438:174–179. doi:10.1016/j.neulet.2008.04.043

86. Shohami E (1999) Dual role of tumor necrosis factor alpha in brain injury. Cytokine Growth Factor Rev 10:119–130. doi:10.1016/S1359-6101(99)00008-8
87. Shohami E, Bass R, Wallach D et al (1996) Inhibition of tumor necrosis factor alpha (TNFalpha) activity in rat brain is associated with cerebroprotection after closed head injury. J Cereb Blood Flow Metab 16:378–384. doi:10.1097/00004647-199605000-00004
88. Folkersma H, Brevé JJP, Tilders FJH et al (2008) Cerebral microdialysis of interleukin (IL)-1beta and IL-6: extraction efficiency and production in the acute phase after severe traumatic brain injury in rats. Acta Neurochir (Wien) 150:1277–1284. doi:10.1007/s00701-008-0151-y, discussion 1284
89. Cui M, Huang Y, Tian C et al (2011) FOXO3a inhibits TNF-α- and IL-1β-induced astrocyte proliferation:Implication for reactive astrogliosis. Glia 59:641–654. doi:10.1002/glia.21134
90. Liberto CM, Albrecht PJ, Herx LM et al (2004) Pro-regenerative properties of cytokine-activated astrocytes. J Neurochem 89:1092–1100. doi:10.1111/j.1471-4159.2004.02420.x
91. Basu A, Krady JK, O'Malley M et al (2002) The Type 1 Interleukin-1 Receptor Is Essential for the Efficient Activation of Microglia and the Induction of Multiple Proinflammatory Mediators in Response to Brain Injury. J Neurosci 22:6071–6082
92. Ching S, He L, Lai W, Quan N (2005) IL-1 type I receptor plays a key role in mediating the recruitment of leukocytes into the central nervous system. Brain Behav Immun 19:127–137. doi:10.1016/j.bbi.2004.06.001
93. Heese K, Hock C, Otten U (2002) Inflammatory Signals Induce Neurotrophin Expression in HumanMicroglialCells.JNeurochem70:699–707.doi:10.1046/j.1471-4159.1998.70020699.x
94. Gabay C, Lamacchia C, Palmer G (2010) IL-1 pathways in inflammation and human diseases. Nat Rev Rheumatol 6:232–241. doi:10.1038/nrrheum.2010.4
95. Smith ED, Prieto GA, Tong L et al (2014) Rapamycin and interleukin-1β impair brain-derived neurotrophic factor-dependent neuron survival by modulating autophagy. J Biol Chem 289:20615–20629. doi:10.1074/jbc.M114.568659
96. Kasai M, Jikoh T, Fukumitsu H, Furukawa S (2014) FGF-2-responsive and spinal cord-resident cells improve locomotor function after spinal cord injury. J Neurotrauma 31:1584–1598. doi:10.1089/neu.2009.1108
97. Lapchak PA, Araujo DM, Carswell S, Hefti F (1993) Distribution of [125I]nerve growth factor in the rat brain following a single intraventricular injection: Correlation with the topographical distribution of trk a messenger RNA-expressing cells. Neuroscience 54:445–460. doi:10.1016/0306-4522(93)90265-H
98. Tong L, Prieto GA, Kramár EA et al (2012) Brain-derived neurotrophic factor-dependent synaptic plasticity is suppressed by interleukin-1β via p38 mitogen-activated protein kinase. J Neurosci 32:17714–17724. doi:10.1523/JNEUROSCI.1253-12.2012
99. Araujo D, Cotman C (1992) Basic FGF in astroglial, microglial, and neuronal cultures: characterization of binding sites and modulation of release by lymphokines and trophic factors. J Neurosci 12:1668–1678
100. Friedman WJ (2005) Interactions of interleukin-1 with neurotrophic factors in the central nervous system: beneficial or detrimental? Mol Neurobiol 32:133–144. doi:10.1385/MN:32:2:133
101. Tong L, Balazs R, Soiampornkul R et al (2008) Interleukin-1 beta impairs brain derived neurotrophic factor-induced signal transduction. Neurobiol Aging 29:1380–1393. doi:10.1016/j.neurobiolaging.2007.02.027
102. van Bruggen N, Thibodeaux H, Palmer JT et al (1999) VEGF antagonism reduces edema formation and tissue damage after ischemia/reperfusion injury in the mouse brain. J Clin Invest 104:1613–1620. doi:10.1172/JCI8218
103. Argaw AT, Asp L, Zhang J et al (2012) Astrocyte-derived VEGF-A drives blood-brain barrier disruption in CNS inflammatory disease. J Clin Invest 122:2454–2468. doi:10.1172/JCI60842
104. Wang H, Wang Y, Li D et al (2015) VEGF inhibits the inflammation in spinal cord injury through activation of autophagy. Biochem Biophys Res Commun 464:453–458. doi:10.1016/j.bbrc.2015.06.146
105. Tolosa L, Mir M, Asensio VJ et al (2008) Vascular endothelial growth factor protects spinal cord motoneurons against glutamate-induced excitotoxicity via phosphatidylinositol 3-kinase. J Neurochem 105:1080–1090. doi:10.1111/j.1471-4159.2007.05206.x

106. Tovar-Y-Romo LB, Tapia R (2010) VEGF protects spinal motor neurons against chronic excitotoxic degeneration in vivo by activation of PI3-K pathway and inhibition of p38MAPK. J Neurochem 115:1090–1101. doi:10.1111/j.1471-4159.2010.06766.x
107. Van Den Bosch L, Storkebaum E, Vleminckx V et al (2004) Effects of vascular endothelial growth factor (VEGF) on motor neuron degeneration. Neurobiol Dis 17:21–28. doi:10.1016/j.nbd.2004.06.004
108. Adeeb N, Mortazavi MM (2014) The role of FGF2 in spinal cord trauma and regeneration research. Brain Behav 4:105–107. doi:10.1002/brb3.207
109. Meng X, Arocena M, Penninger J et al (2011) PI3K mediated electrotaxis of embryonic and adult neural progenitor cells in the presence of growth factors. Exp Neurol 227:210–217. doi:10.1016/j.expneurol.2010.11.002
110. Johnson-Farley NN, Patel K, Kim D, Cowen DS (2007) Interaction of FGF-2 with IGF-1 and BDNF in stimulating Akt, ERK, and neuronal survival in hippocampal cultures. Brain Res 1154:40–49. doi:10.1016/j.brainres.2007.04.026
111. Jakovcevski I, Wu J, Karl N et al (2007) Glial scar expression of CHL1, the close homolog of the adhesion molecule L1, limits recovery after spinal cord injury. J Neurosci 27:7222–7233. doi:10.1523/JNEUROSCI.0739-07.2007
112. Hashimoto M, Sagara Y, Langford D et al (2002) Fibroblast growth factor 1 regulates signaling via the glycogen synthase kinase-3beta pathway. Implications for neuroprotection. J Biol Chem 277:32985–32991. doi:10.1074/jbc.M202803200
113. Tsai M-C, Shen L-F, Kuo H-S et al (2008) Involvement of acidic fibroblast growth factor in spinal cord injury repair processes revealed by a proteomics approach. Mol Cell Proteomics 7:1668–1687. doi:10.1074/mcp.M800076-MCP200
114. Rottlaender A, Villwock H, Addicks K, Kuerten S (2011) Neuroprotective role of fibroblast growth factor-2 in experimental autoimmune encephalomyelitis. Immunology 133:370–378. doi:10.1111/j.1365-2567.2011.03450.x
115. Ruffini F, Furlan R, Poliani PL et al (2001) Fibroblast growth factor-II gene therapy reverts the clinical course and the pathological signs of chronic experimental autoimmune encephalomyelitis in C57BL/6 mice. Gene Ther 8:1207
116. Jones M (2000) Heparan sulfate proteoglycan isoforms of the CD44 hyaluronan receptor induced in human inflammatory macrophages can function as paracrine regulators of fibroblast growth factor action. J Biol Chem 275:7964–7974. doi:10.1074/jbc.275.11.7964
117. Pedersen MØ, Larsen A, Pedersen DS et al (2009) Metallic gold treatment reduces proliferation of inflammatory cells, increases expression of VEGF and FGF, and stimulates cell proliferation in the subventricular zone following experimental traumatic brain injury. Histol Histopathol 24:573–586
118. Ghirnikar RS, Lee YL, Eng LF (1998) Inflammation in traumatic brain injury: role of cytokines and chemokines. Neurochem Res 23:329–340
119. Reilly JF, Maher PA, Kumari VG (1998) Regulation of astrocyte GFAP expression by TGF-beta1 and FGF-2. Glia 22:202–210
120. Goddard DR, Berry M, Kirvell SL, Butt AM (2002) Fibroblast growth factor-2 induces astroglial and microglial reactivity in vivo. J Anat 200:57–67. doi:10.1046/j.0021-8782.2001.00002.x
121. Lee TT, Green BA, Dietrich WD, Yezierski RP (1999) Neuroprotective effects of basic fibroblast growth factor following spinal cord contusion injury in the rat. J Neurotrauma 16:347–356. doi:10.1089/neu.1999.16.347
122. Rabchevsky AG, Fugaccia I, Turner AF et al (2000) Basic fibroblast growth factor (bFGF) enhances functional recovery following severe spinal cord injury to the rat. Exp Neurol 164:280–291. doi:10.1006/exnr.2000.7399
123. Jiang Y, Wei N, Lu T et al (2011) Intranasal brain-derived neurotrophic factor protects brain from ischemic insult via modulating local inflammation in rats. Neuroscience 172:398–405
124. Bovolenta R, Zucchini S, Paradiso B et al (2010) Hippocampal FGF-2 and BDNF overexpression attenuates epileptogenesis-associated neuroinflammation and reduces spontaneous recurrent seizures. J Neuroinflammation 7:81

125. Schäbitz W-R, Sommer C, Zoder W et al (2000) Intravenous brain-derived neurotrophic factor reduces infarct size and counterregulates Bax and Bcl-2 expression after temporary focal cerebral ischemia. Stroke 31:2212–2217
126. Sharma HS (2006) Post-traumatic application of brain-derived neurotrophic factor and glia-derived neurotrophic factor on the rat spinal cord enhances neuroprotection and improves motor function. Acta Neurochir Suppl 96:329–334. doi:10.1007/3-211-30714-1-69
127. Sharma HS (2007) A select combination of neurotrophins enhances neuroprotection and functional recovery following spinal cord injury. Ann N Y Acad Sci 1122:95–111. doi:10.1196/annals.1403.007
128. Ghirnikar RS, Lee YL, Eng LF (2000) Chemokine antagonist infusion attenuates cellular infiltration following spinal cord contusion injury in rat. J Neurosci Res 59:63–73
129. Lüttichau HR, Stine J, Boesen TP et al (2000) A highly selective Cc chemokine receptor (Ccr)8 antagonist encoded by the poxvirus molluscum contagiosum. J Exp Med 191:171–180. doi:10.1084/jem.191.1.171
130. Mantovani A, Bonecchi R, Locati M (2006) Tuning inflammation and immunity by chemokine sequestration: decoys and more. Nat Rev Immunol 6:907–918. doi:10.1038/nri1964
131. Ghirnikar R, Lee Y, Li J, Eng L (1998) Chemokine inhibition in rat stab wound brain injury using antisense oligodeoxynucleotides. Neurosci Lett 247:21–24. doi:10.1016/S0304-3940(98)00268-7
132. Chen M-H, Ren Q-X, Yang W-F et al (2013) Influences of HIF-lα on Bax/Bcl-2 and VEGF expressions in rats with spinal cord injury. Int J Clin Exp Pathol 6:2312–2322
133. Mott JL, Kobayashi S, Bronk SF, Gores GJ (2007) mir-29 regulates Mcl-1 protein expression and apoptosis. Oncogene 26:6133–6140. doi:10.1038/sj.onc.1210436
134. Liu X-J, Zheng X-P, Zhang R et al (2015) Combinatorial effects of miR-20a and miR-29b on neuronal apoptosis induced by spinal cord injury. Int J Clin Exp Pathol 8:3811–3818
135. Hyun J, Grova M, Nejadnik H et al (2013) Enhancing in vivo survival of adipose-derived stromal cells through Bcl-2 overexpression using a minicircle vector. Stem Cells Transl Med 2:690
136. Pannu R, Christie DK, Barbosa E et al (2007) Post-trauma Lipitor treatment prevents endothelial dysfunction, facilitates neuroprotection, and promotes locomotor recovery following spinal cord injury. J Neurochem 101:182–200. doi:10.1111/j.1471-4159.2006.04354.x
137. Iannotti CA, Clark M, Horn KP et al (2011) A combination immunomodulatory treatment promotes neuroprotection and locomotor recovery after contusion SCI. Exp Neurol 230:3–15. doi:10.1016/j.expneurol.2010.03.010
138. Abrams MB, Nilsson I, Lewandowski SA et al (2012) Imatinib enhances functional outcome after spinal cord injury. PLoS One 7, e38760. doi:10.1371/journal.pone.0038760
139. Schaal SM, Garg M Sen, Ghosh M et al (2012) The therapeutic profile of rolipram, PDE target and mechanism of action as a neuroprotectant following spinal cord injury. PLoS One 7:e43634. doi:10.1371/journal.pone.0043634
140. Lee KD, Chow WN, Sato-Bigbee C et al (2009) FTY720 reduces inflammation and promotes functional recovery after spinal cord injury. J Neurotrauma 26:2335–2344. doi:10.1089/neu.2008.0840
141. Paniagua RT, Sharpe O, Ho PP et al (2006) Selective tyrosine kinase inhibition by imatinib mesylate for the treatment of autoimmune arthritis. J Clin Invest 116:2633–2642. doi:10.1172/JCI28546
142. Su EJ, Fredriksson L, Kanzawa M et al (2015) Imatinib treatment reduces brain injury in a murine model of traumatic brain injury. Front Cell Neurosci 9:385. doi:10.3389/fncel.2015.00385
143. Zhan Y, Krafft PR, Lekic T et al (2015) Imatinib preserves blood-brain barrier integrity following experimental subarachnoid hemorrhage in rats. J Neurosci Res 93:94–103. doi:10.1002/jnr.23475
144. Buttini M, Mir A, Appel K et al (1997) Lipopolysaccharide induces expression of tumour necrosis factor alpha in rat brain: inhibition by methylprednisolone and by rolipram. Br J Pharmacol 122:1483–1489. doi:10.1038/sj.bjp.0701502

145. Yoshikawa M, Suzumura A, Tamaru T et al (1999) Effects of phosphodiesterase inhibitors on cytokine production by microglia. Mult Scler 5:126–133. doi:10.1177/135245859900500210
146. Zhu J, Mix E, Winblad B (2006) The antidepressant and antiinflammatory effects of rolipram in the central nervous system. CNS Drug Rev 7:387–398. doi:10.1111/j.1527-3458.2001.tb00206.x
147. Beaumont E, Whitaker CM, Burke DA et al (2009) Effects of rolipram on adult rat oligodendrocytes and functional recovery after contusive cervical spinal cord injury. Neuroscience 163:985–990. doi:10.1016/j.neuroscience.2009.07.039
148. Whitaker CM, Beaumont E, Wells MJ et al (2008) Rolipram attenuates acute oligodendrocyte death in the adult rat ventrolateral funiculus following contusive cervical spinal cord injury. Neurosci Lett 438:200–204. doi:10.1016/j.neulet.2008.03.087
149. Hätinen S, Sairanen M, Sirviö J, Jolkkonen J (2008) Improved sensorimotor function by rolipram following focal cerebral ischemia in rats. Restor Neurol Neurosci 26:493–499
150. Sasaki T, Kitagawa K, Omura-Matsuoka E et al (2007) The phosphodiesterase inhibitor rolipram promotes survival of newborn hippocampal neurons after ischemia. Stroke 38:1597–1605. doi:10.1161/STROKEAHA.106.476754
151. Atkins CM, Kang Y, Furones C et al (2012) Postinjury treatment with rolipram increases hemorrhage after traumatic brain injury. J Neurosci Res 90:1861–1871. doi:10.1002/jnr.23069
152. Atkins CM, Cepero ML, Kang Y et al (2013) Effects of early rolipram treatment on histopathological outcome after controlled cortical impact injury in mice. Neurosci Lett 532:1–6. doi:10.1016/j.neulet.2012.10.019
153. Koopmans GC, Deumens R, Buss A et al (2009) Acute rolipram/thalidomide treatment improves tissue sparing and locomotion after experimental spinal cord injury. Exp Neurol 216:490–498. doi:10.1016/j.expneurol.2009.01.005
154. George A, Marziniak M, Schäfers M et al (2000) Thalidomide treatment in chronic constrictive neuropathy decreases endoneurial tumor necrosis factor-α, increases interleukin-10 and has long-term effects on spinal cord dorsal horn met-enkephalin. Pain 88:267–275. doi:10.1016/S0304-3959(00)00333-X
155. Wang Y, Gu J, Wang J et al (2012) BDNF and NT-3 expression by using glucocorticoid-induced bicistronic expression vector pGC-BDNF-IRES-NT3 protects apoptotic cells in a cellular injury model. Brain Res 1448:137–143. doi:10.1016/j.brainres.2012.02.007
156. Hyakkoku K, Nakajima Y, Izuta H et al (2009) Thalidomide protects against ischemic neuronal damage induced by focal cerebral ischemia in mice. Neuroscience 159:760–769. doi:10.1016/j.neuroscience.2008.12.043
157. Papa S, Caron I, Erba E et al (2016) Early modulation of pro-inflammatory microglia by minocycline loaded nanoparticles confers long lasting protection after spinal cord injury. Biomaterials 75:13–24. doi:10.1016/j.biomaterials.2015.10.015
158. Sanchez Mejia RO, Ona VO, Li M, Friedlander RM (2001) Minocycline reduces traumatic brain injury-mediated caspase-1 activation, tissue damage, and neurological dysfunction. Neurosurgery 48:1393–1401. doi:10.1227/00006123-200106000-00051
159. Bye N, Habgood MD, Callaway JK et al (2007) Transient neuroprotection by minocycline following traumatic brain injury is associated with attenuated microglial activation but no changes in cell apoptosis or neutrophil infiltration. Exp Neurol 204:220–233. doi:10.1016/j.expneurol.2006.10.013
160. Wells JEA, Hurlbert RJ, Fehlings MG, Yong VW (2003) Neuroprotection by minocycline facilitates significant recovery from spinal cord injury in mice. Brain 126:1628–1637. doi:10.1093/brain/awg178
161. Lee SM, Yune TY, Kim SJ et al (2003) Minocycline reduces cell death and improves functional recovery after traumatic spinal cord injury in the rat. J Neurotrauma 20:1017–1027. doi:10.1089/089771503770195867
162. Homsi S, Federico F, Croci N et al (2009) Minocycline effects on cerebral edema: relations with inflammatory and oxidative stress markers following traumatic brain injury in mice. Brain Res 1291:122–132. doi:10.1016/j.brainres.2009.07.031
163. Kelso ML, Scheff NN, Scheff SW, Pauly JR (2011) Melatonin and minocycline for combinatorial therapy to improve functional and histopathological deficits following traumatic brain injury. Neurosci Lett 488:60–64. doi:10.1016/j.neulet.2010.11.003

164. Arvin KL, Han BH, Du Y et al (2002) Minocycline markedly protects the neonatal brain against hypoxic-ischemic injury. Ann Neurol 52:54–61. doi:10.1002/ana.10242
165. Wayne State University Safety and Feasibility of Minocycline in the Treatment of Traumatic Brain Injury (2014) In: ClinicalTrials.gov [Internet]. National Library of Medicine, Bethesda. https://clinicaltrials.gove/ct2/show/NCT01058395NLMIde
166. Kovesdi E, Kamnaksh A, Wingo D et al (2012) Acute minocycline treatment mitigates the symptoms of mild blast-induced traumatic brain injury. Front Neurol. doi:10.3389/fneur.2012.00111
167. O'Collins VE, Macleod MR, Donnan GA et al (2006) 1,026 experimental treatments in acute stroke. Ann Neurol 59:467–477. doi:10.1002/ana.20741
168. Cobbs CS, Chen J, Greenberg DA, Graham SH (1998) Vascular endothelial growth factor expression in transient focal cerebral ischemia in the rat. Neurosci Lett 249:79–82. doi:10.1016/S0304-3940(98)00377-2
169. Krupinski J, Kaluza J, Kumar P et al (1994) Role of angiogenesis in patients with cerebral ischemic stroke. Stroke 25:1794–1798. doi:10.1161/01.STR.25.9.1794
170. Brogi E, Wu T, Namiki A, Isner JM (1994) Indirect angiogenic cytokines upregulate VEGF and bFGF gene expression in vascular smooth muscle cells, whereas hypoxia upregulates VEGF expression only. Circulation 90:649–652. doi:10.1161/01.CIR.90.2.649
171. Yang J-P, Liu X-F, Liu H-J et al (2008) Extracellular signal-regulated kinase involved in NGF/VEGF-induced neuroprotective effect. Neurosci Lett 434:212–217. doi:10.1016/j.neulet.2008.01.078
172. Yang J-P, Liu H-J, Li Y (2009) Effect of Endoplasmic Reticulum Stress in VEGF-Induced Neuroprotection. J. Investig, Surg
173. Zhu W, Mao Y, Zhao Y, et al. (2005) Transplantation of Vascular Endothelial Growth Factor-transfected Neural Stem Cells into the Rat Brain Provides Neuroprotection after Transient Focal Cerebral Ischemia. Neurosurgery 57
174. Chu K, Park K-I, Lee S-T et al (2005) Combined treatment of vascular endothelial growth factor and human neural stem cells in experimental focal cerebral ischemia. Neurosci Res 53:384–390. doi:10.1016/j.neures.2005.08.010
175. Kaya D, Gürsoy-Özdemir Y, Yemisci M et al (2005) VEGF protects brain against focal ischemia without increasing blood–brain permeability when administered intracerebroventricularly. J Cereb Blood Flow 25:1111–1118. doi:10.1038/sj.jcbfm.9600109
176. Lutton C, Young YW, Williams R et al (2012) Combined VEGF and PDGF treatment reduces secondary degeneration after spinal cord injury. J Neurotrauma 29:957–970. doi:10.1089/neu.2010.1423
177. Kim HM, Hwang DH, Lee JE et al (2009) Ex vivo VEGF delivery by neural stem cells enhances proliferation of glial progenitors, angiogenesis, and tissue sparing after spinal cord injury. PLoS One 4, e4987. doi:10.1371/journal.pone.0004987
178. Hicks RR, Martin VB, Zhang L, Seroogy KB (1999) Mild experimental brain injury differentially alters the expression of neurotrophin and neurotrophin receptor mRNAs in the hippocampus. Exp Neurol 160:469–478. doi:10.1006/exnr.1999.7216
179. De Laporte L, des Rieux A, Tuinstra HM, et al. (2011) Vascular endothelial growth factor and fibroblast growth factor 2 delivery from spinal cord bridges to enhance angiogenesis following injury. J Biomed Mater Res Part A 98A:372–382. doi:10.1002/jbm.a.33112
180. Liu Y, Figley S, Spratt SK et al (2010) An engineered transcription factor which activates VEGF-A enhances recovery after spinal cord injury. Neurobiol Dis 37:384–393
181. Nori S, Okada Y, Yasuda A et al (2011) Grafted human-induced pluripotent stem-cell–derived neurospheres promote motor functional recovery after spinal cord injury in mice. Proc Natl Acad Sci 108:16825–16830
182. Nakajima H, Uchida K, Yayama T et al (2010) Targeted retrograde gene delivery of brain-derived neurotrophic factor suppresses apoptosis of neurons and oligodendroglia after spinal cord injury in rats. Spine (Phila Pa 1976) 35:497–504

183. Hicks R, Numan S, Dhillon H et al (1997) Alterations in BDNF and NT-3 mRNAs in rat hippocampus after experimental brain trauma. Mol Brain Res 48:401–406. doi:10.1016/S0169-328X(97)00158-7
184. Novikova LN, Novikov LN, Kellerth JO (2000) BDNF abolishes the survival effect of NT-3 in axotomized Clarke neurons of adult rats. J Comp Neurol 428:671–680
185. Bradbury EJ, King VR, Simmons LJ et al (1998) NT-3, but not BDNF, prevents atrophy and death of axotomized spinal cord projection neurons. Eur J Neurosci 10:3058–3068. doi:10.1046/j.1460-9568.1998.00307.x
186. Uchida K, Nakajima H, Inukai T et al (2008) Adenovirus-mediated retrograde transfer of neurotrophin-3 gene enhances survival of anterior horn neurons of twy/twy mice with chronic mechanical compression of the spinal cord. J Neurosci Res 86:1789–1800. doi:10.1002/jnr.21627
187. Volpato FZ, Führmann T, Migliaresi C et al (2013) Using extracellular matrix for regenerative medicine in the spinal cord. Biomaterials 34:4945–4955. doi:10.1016/j.biomaterials.2013.03.057
188. Zhou L, Baumgartner BJ, Hill-Felberg SJ et al (2003) Neurotrophin-3 expressed in situ induces axonal plasticity in the adult injured spinal cord. J Neurosci 23:1424–1431
189. Hou S, Nicholson L, van Niekerk E et al (2012) Dependence of regenerated sensory axons on continuous neurotrophin-3 delivery. J Neurosci 32:13206–13220. doi:10.1523/JNEUROSCI.5041-11.2012
190. Blits B, Kitay BM, Farahvar A et al (2005) Lentiviral vector-mediated transduction of neural progenitor cells before implantation into injured spinal cord and brain to detect their migration, deliver neurotrophic factors and repair tissue. Restor Neurol Neurosci 23:313–324
191. Tuszynski MH, Grill R, Jones LL et al (2003) NT-3 gene delivery elicits growth of chronically injured corticospinal axons and modestly improves functional deficits after chronic scar resection. Exp Neurol 181:47–56. doi:10.1016/S0014-4886(02)00055-9
192. Taylor LL (2006) Neurotrophin-3 gradients established by lentiviral gene delivery promote short-distance axonal bridging beyond cellular grafts in the injured spinal cord. J Neurosci 26:9713–9721. doi:10.1523/JNEUROSCI.0734-06.2006
193. Schnell L, Schneider R, Kolbeck R et al (1994) Neurotrophin-3 enhances sprouting of corticospinal tract during development and after adult spinal cord lesion. Nature 367:170–173. doi:10.1038/367170a0
194. Kusano K, Enomoto M, Hirai T et al (2010) Transplanted neural progenitor cells expressing mutant NT3 promote myelination and partial hindlimb recovery in the chronic phase after spinal cord injury. Biochem Biophys Res Commun 393:812–817. doi:10.1016/j.bbrc.2010.02.088
195. Bonner JF, Blesch A, Neuhuber B, Fischer I (2010) Promoting directional axon growth from neural progenitors grafted into the injured spinal cord. J Neurosci Res 88:1182–1192. doi:10.1002/jnr.22288
196. Blesch A, Tuszynski MH (2007) Transient growth factor delivery sustains regenerated axons after spinal cord injury. J Neurosci 27:10535–10545. doi:10.1523/JNEUROSCI.1903-07.2007
197. Tobias C, Shumsky J, Shibata M et al (2003) Delayed grafting of BDNF and NT-3 producing fibroblasts into the injured spinal cord stimulates sprouting, partially rescues axotomized red nucleus neurons from loss and atrophy, and provides limited regeneration. Exp Neurol 184:97–113. doi:10.1016/S0014-4886(03)00394-7
198. Sasaki MM (2009) BDNF-hypersecreting human mesenchymal stem cells promote functional recovery, axonal sprouting, and protection of corticospinal neurons after spinal cord injury. J Neurosci 29:14932–14941. doi:10.1523/JNEUROSCI.2769-09.2009
199. Nakajima H, Uchida K, Kobayashi S et al (2007) Rescue of rat anterior horn neurons after spinal cord injury by retrograde transfection of adenovirus vector carrying brain-derived neurotrophic factor gene. J Neurotrauma 24:703–712. doi:10.1089/neu.2006.0004
200. Koda M, Murakami M, Ino H et al (2002) Brain-derived neurotrophic factor suppresses delayed apoptosis of oligodendrocytes after spinal cord injury in rats. J Neurotrauma 19:777–785. doi:10.1089/08977150260139147

201. Gao L, Díaz-Corrales FJ, Carrillo F et al (2010) Brain-derived neurotrophic factor G196A polymorphism and clinical features in Parkinson's disease. Acta Neurol Scand 122:41–45. doi:10.1111/j.1600-0404.2009.01253.x
202. Hyman C, Hofer M, Barde YA et al (1991) BDNF is a neurotrophic factor for dopaminergic neurons of the substantia nigra. Nature 350:230–232. doi:10.1038/350230a0
203. Zhang S, Liu X, Liu Z et al (2009) Stem cells modified by brain-derived neurotrophic factor to promote stem cells differentiation into neurons and enhance neuromotor function after brain injury. Chinese J Traumatol English Ed 12:195–199. doi:10.3760/cma.j.issn.1008-1275.2009.04.001
204. Zhang Y, Pardridge WM (2006) Blood-brain barrier targeting of BDNF improves motor function in rats with middle cerebral artery occlusion. Brain Res 1111:227–229. doi:10.1016/j.brainres.2006.07.005
205. Pardridge WM (2002) Blood-brain barrier drug targeting enables neuroprotection in brain ischemia following delayed intravenous administration of neurotrophins. In: Alzheimer C (ed) Neuroprotection CNS. Mol Cell Biol, Springer, Boston, pp 397–430
206. Ma H, Yu B, Kong L et al (2012) Neural stem cells over-expressing brain-derived neurotrophic factor (BDNF) stimulate synaptic protein expression and promote functional recovery following transplantation in rat model of traumatic brain injury. Neurochem Res 37:69–83. doi:10.1007/s11064-011-0584-1
207. Blits B, Oudega M, Boer G et al (2003) Adeno-associated viral vector-mediated neurotrophin gene transfer in the injured adult rat spinal cord improves hind-limb function. Neuroscience 118:271–281. doi:10.1016/S0306-4522(02)00970-3
208. Xu XM, Guénard V, Kleitman N et al (1995) A combination of BDNF and NT-3 promotes supraspinal axonal regeneration into Schwann cell grafts in adult rat thoracic spinal cord. Exp Neurol 134:261–272. doi:10.1006/exnr.1995.1056
209. Bamber NI, Li H, Lu X et al (2001) Neurotrophins BDNF and NT-3 promote axonal re-entry into the distal host spinal cord through Schwann cell-seeded mini-channels. Eur J Neurosci 13:257–268. doi:10.1111/j.1460-9568.2001.01387.x
210. Xu XM, Zhang S-X, Li H et al (1999) Regrowth of axons into the distal spinal cord through a Schwann-cell-seeded mini-channel implanted into hemisected adult rat spinal cord. Eur J Neurosci 11:1723–1740. doi:10.1046/j.1460-9568.1999.00591.x
211. Azim K, Raineteau O, Butt AM (2012) Intraventricular injection of FGF-2 promotes generation of oligodendrocyte-lineage cells in the postnatal and adult forebrain. Glia 60:1977–1990. doi:10.1002/glia.22413
212. Bansal R, Lakhina V, Remedios R, Tole S (2003) Expression of FGF receptors 1, 2, 3 in the embryonic and postnatal mouse brain compared with Pdgfralpha, Olig2 and Plp/dm20: implications for oligodendrocyte development. Dev Neurosci 25:83–95.
213. Timmer M, Cesnulevicius K, Winkler C et al (2007) Fibroblast growth factor (FGF)-2 and FGF receptor 3 are required for the development of the substantia nigra, and FGF-2 plays a crucial role for the rescue of dopaminergic neurons after 6-hydroxydopamine lesion. J Neurosci 27:459–471. doi:10.1523/JNEUROSCI.4493-06.2007
214. Butt AM, Dinsdale J (2015) Opposing actions of fibroblast growth factor-2 on early and late oligodendrocyte lineage cells in vivo. J Neuroimmunol 166:75–87. doi:10.1016/j.jneuroim.2005.05.015
215. Tabata Y (2009) Biomaterial technology for tissue engineering applications. J R Soc Interface 6(Suppl 3):S311–S324. doi:10.1098/rsif.2008.0448.focus
216. Rooney GE, Vaishya S, Ameenuddin S et al (2008) Rigid fixation of the spinal column improves scaffold alignment and prevents scoliosis in the transected rat spinal cord. Spine (Phila Pa 1976) 33:E914–E919. doi:10.1097/BRS.0b013e318186b2b1
217. Tasdemiroglu E, Tibbs PA (1995) Long-term follow-up results of thoracolumbar fractures after posterior instrumentation. Spine (Phila. Pa. 1976). 20:1704–1708
218. Wang J, Zheng J, Zheng Q et al (2015) FGL-functionalized self-assembling nanofiber hydrogel as a scaffold for spinal cord-derived neural stem cells. Mater Sci Eng C Mater Biol Appl 46:140–147. doi:10.1016/j.msec.2014.10.019

219. Kubinová S, Syková E (2010) Nanotechnologies in regenerative medicine. Minim Invasive Ther Allied Technol 19:144–156. doi:10.3109/13645706.2010.481398
220. Madigan NN, McMahon S, O'Brien T et al (2009) Current tissue engineering and novel therapeutic approaches to axonal regeneration following spinal cord injury using polymer scaffolds. Respir Physiol Neurobiol 169:183–199. doi:10.1016/j.resp.2009.08.015
221. Wu M, Yuan H, Li K, Qiu D-L (2015) Cellular transplantation-based evolving treatment options in spinal cord injury. Cell Biochem Biophys 71:1–8. doi:10.1007/s12013-014-0174-3
222. Perale G, Rossi F, Sundstrom E et al (2011) Hydrogels in spinal cord injury repair strategies. ACS Chem Neurosci 2:336–345. doi:10.1021/cn200030w
223. Aurand ER, Lampe KJ, Bjugstad KB (2011) Defining and designing polymers and hydrogels for neural tissue engineering. Neurosci Res 72:199–213. doi:10.1016/j.neures.2011.12.005
224. Tsai EC, Dalton PD, Shoichet MS, Tator CH (2006) Matrix inclusion within synthetic hydrogel guidance channels improves specific supraspinal and local axonal regeneration after complete spinal cord transection. Biomaterials 27:519–533. doi:10.1016/j.biomaterials.2005.07.025
225. Woerly S, Petrov P, Syková E et al (1999) Neural tissue formation within porous hydrogels implanted in brain and spinal cord lesions: ultrastructural, immunohistochemical, and diffusion studies. Tissue Eng 5:467–488. doi:10.1089/ten.1999.5.467
226. Straley KS, Foo CWP, Heilshorn SC (2010) Biomaterial design strategies for the treatment of spinal cord injuries. J Neurotrauma 27:1–19. doi:10.1089/neu.2009.0948
227. Hynes SR, Rauch MF, Bertram JP, Lavik EB (2009) A library of tunable poly(ethylene glycol)/poly(L-lysine) hydrogels to investigate the material cues that influence neural stem cell differentiation. J Biomed Mater Res A 89:499–509. doi:10.1002/jbm.a.31987
228. Dillon GP, Sridharan A, Ranieri JP, Bellamkonda RV (1998) The influence of physical structure and charge on neurite extension in a 3D hydrogel scaffold. J Biomater Sci Polym Ed 9:1049–1069. doi:10.1163/156856298X00325
229. Varon S (1979) The culture of chick embryo dorsal root ganglionic cells on polylysine-coated plastic. Neurochem Res 4:155–173. doi:10.1007/BF00964141
230. Crompton KE, Goud JD, Bellamkonda RV et al (2007) Polylysine-functionalised thermoresponsive chitosan hydrogel for neural tissue engineering. Biomaterials 28:441–449. doi:10.1016/j.biomaterials.2006.08.044
231. Dadsetan M, Knight AM, Lu L et al (2009) Stimulation of neurite outgrowth using positively charged hydrogels. Biomaterials 30:3874–3881. doi:10.1016/j.biomaterials.2009.04.018
232. Young T-H, Tu H-R, Chan C-C et al (2009) The enhancement of dermal papilla cell aggregation by extracellular matrix proteins through effects on cell-substratum adhesivity and cell motility. Biomaterials 30:5031–5040. doi:10.1016/j.biomaterials.2009.05.065
233. Ryan PL, Foty RA, Kohn J, Steinberg MS (2001) Tissue spreading on implantable substrates is a competitive outcome of cell-cell vs. cell-substratum adhesivity. Proc Natl Acad Sci U S A 98:4323–4327. doi:10.1073/pnas.071615398
234. Young T-H, Lee C-Y, Chiu H-C et al (2008) Self-assembly of dermal papilla cells into inductive spheroidal microtissues on poly(ethylene-co-vinyl alcohol) membranes for hair follicle regeneration. Biomaterials 29:3521–3530. doi:10.1016/j.biomaterials.2008.05.013
235. Kundu AK, Khatiwala CB, Putnam AJ (2009) Extracellular matrix remodeling, integrin expression, and downstream signaling pathways influence the osteogenic differentiation of mesenchymal stem cells on poly(lactide-co-glycolide) substrates. Tissue Eng Part A 15:273–283. doi:10.1089/ten.tea.2008.0055
236. Liu JF, Chen YM, Yang JJ et al (2011) Dynamic behavior and spontaneous differentiation of mouse embryoid bodies on hydrogel substrates of different surface charge and chemical structures. Tissue Eng Part A 17:2343–2357. doi:10.1089/ten.TEA.2011.0034
237. Leach JB, Brown XQ, Jacot JG et al (2007) Neurite outgrowth and branching of PC12 cells on very soft substrates sharply decreases below a threshold of substrate rigidity. J Neural Eng 4:26–34. doi:10.1088/1741-2560/4/2/003
238. Gunn JW, Turner SD, Mann BK (2005) Adhesive and mechanical properties of hydrogels influence neurite extension. J Biomed Mater Res A 72:91–97. doi:10.1002/jbm.a.30203

239. Balgude A (2001) Agarose gel stiffness determines rate of DRG neurite extension in 3D cultures. Biomaterials 22:1077–1084. doi:10.1016/S0142-9612(00)00350-1
240. Engler A, Bacakova L, Newman C et al (2004) Substrate compliance versus ligand density in cell on gel responses. Biophys J 86:617–628. doi:10.1016/S0006-3495(04)74140-5
241. Thomas TW, DiMilla PA (2000) Spreading and motility of human glioblastoma cells on sheets of silicone rubber depend on substratum compliance. Med Biol Eng Comput 38:360–370. doi:10.1007/BF02347059
242. Lampe KJ, Antaris AL, Heilshorn SC (2013) Design of three-dimensional engineered protein hydrogels for tailored control of neurite growth. Acta Biomater 9:5590–5599. doi:10.1016/j.actbio.2012.10.033
243. Ribeiro A, Vargo S, Powell EM, Leach JB (2012) Substrate three-dimensionality induces elemental morphological transformation of sensory neurons on a physiologic timescale. Tissue Eng Part A 18:93–102
244. Lampe KJ, Mooney RG, Bjugstad KB, Mahoney MJ (2010) Effect of macromer weight percent on neural cell growth in 2D and 3D nondegradable PEG hydrogel culture. J Biomed Mater Res A 94:1162–1171. doi:10.1002/jbm.a.32787
245. Cukierman E, Pankov R, Yamada KM (2002) Cell interactions with three-dimensional matrices. Curr Opin Cell Biol 14:633–640. doi:10.1016/S0955-0674(02)00364-2
246. Fraley SI, Feng Y, Krishnamurthy R et al (2010) A distinctive role for focal adhesion proteins in three-dimensional cell motility. Nat Cell Biol 12:598–604. doi:10.1038/ncb2062
247. Kubow KE, Horwitz AR (2011) Reducing background fluorescence reveals adhesions in 3D matrices. Nat Cell Biol 13:3–5. doi:10.1038/ncb0111-3, author reply 5–7
248. Wang L-S, Chung JE, Chan PP-Y, Kurisawa M (2010) Injectable biodegradable hydrogels with tunable mechanical properties for the stimulation of neurogenesic differentiation of human mesenchymal stem cells in 3D culture. Biomaterials 31:1148–1157. doi:10.1016/j.biomaterials.2009.10.042
249. Engler AJ, Sen S, Sweeney HL, Discher DE (2006) Matrix elasticity directs stem cell lineage specification. Cell 126:677–689. doi:10.1016/j.cell.2006.06.044
250. Tse JR, Engler AJ (2011) Stiffness gradients mimicking in vivo tissue variation regulate mesenchymal stem cell fate. PLoS One 6, e15978. doi:10.1371/journal.pone.0015978
251. Saha K, Keung AJ, Irwin EF et al (2008) Substrate modulus directs neural stem cell behavior. Biophys J 95:4426–4438. doi:10.1529/biophysj.108.132217
252. Banerjee A, Arha M, Choudhary S et al (2009) The influence of hydrogel modulus on the proliferation and differentiation of encapsulated neural stem cells. Biomaterials 30:4695–4699. doi:10.1016/j.biomaterials.2009.05.050
253. Gelain F, Horii A, Zhang S (2007) Designer self-assembling peptide scaffolds for 3-d tissue cell cultures and regenerative medicine. Macromol Biosci 7:544–551. doi:10.1002/mabi.200700033
254. Liu Y, Ye H, Satkunendrarajah K et al (2013) A self-assembling peptide reduces glial scarring, attenuates post-traumatic inflammation and promotes neurological recovery following spinal cord injury. Acta Biomater 9:8075–8088. doi:10.1016/j.actbio.2013.06.001
255. Fändrich M, Fletcher MA, Dobson CM (2001) Amyloid fibrils from muscle myoglobin. Nature 410:165–166. doi:10.1038/35065514
256. Lynn DG, Meredith SC (2000) Review: model peptides and the physicochemical approach to beta-amyloids. J Struct Biol 130:153–173. doi:10.1006/jsbi.2000.4287
257. Shtilerman MD, Ding TT, Lansbury PT (2002) Molecular Crowding Accelerates Fibrillization of α-Synuclein: Could an Increase in the Cytoplasmic Protein Concentration Induce Parkinson's Disease? †. Biochemistry 41:3855–3860. doi:10.1021/bi0120906
258. Reches M, Porat Y, Gazit E (2002) Amyloid fibril formation by pentapeptide and tetrapeptide fragments of human calcitonin. J Biol Chem 277:35475–35480. doi:10.1074/jbc.M206039200
259. Caldeira J, Silva AM et al (2015) Microenvironment-Inspired Materials For Tissue Regeneration. http://www.esciencecentral.org/ebooks/biomimetic-approaches-for-tissue-healing/microenvironmentinspired-materials-for-tissue-regeneration.php. Accessed 3 Mar 2016

260. Ellis-Behnke RG, Liang Y-X, You S-W et al (2006) Nano neuro knitting: peptide nanofiber scaffold for brain repair and axon regeneration with functional return of vision. Proc Natl Acad Sci U S A 103:5054–5059. doi:10.1073/pnas.0600559103
261. Guo J, Leung KKG, Su H et al (2009) Self-assembling peptide nanofiber scaffold promotes the reconstruction of acutely injured brain. Nanomedicine 5:345–351. doi:10.1016/j.nano.2008.12.001
262. Silva GA, Czeisler C, Niece KL et al (2004) Selective differentiation of neural progenitor cells by high-epitope density nanofibers. Science 303:1352–1355. doi:10.1126/science.1093783
263. Tysseling-Mattiace VM, Sahni V, Niece KL et al (2008) Self-assembling nanofibers inhibit glial scar formation and promote axon elongation after spinal cord injury. J Neurosci 28:3814–3823. doi:10.1523/JNEUROSCI.0143-08.2008
264. Tysseling VM, Sahni V, Pashuck ET et al (2010) Self-assembling peptide amphiphile promotes plasticity of serotonergic fibers following spinal cord injury. J Neurosci Res 88:3161–3170. doi:10.1002/jnr.22472
265. Guo J, Su H, Zeng Y et al (2007) Reknitting the injured spinal cord by self-assembling peptide nanofiber scaffold. Nanomedicine 3:311–321. doi:10.1016/j.nano.2007.09.003
266. Jiang T, Carbone EJ, Lo KW-H, Laurencin CT (2014) Electrospinning of polymer nanofibers for tissue regeneration. Prog Polym Sci 46:1–24. doi:10.1016/j.progpolymsci.2014.12.001
267. Niece KL, Hartgerink JD, Donners JJJM, Stupp SI (2003) Self-assembly combining two bioactive peptide-amphiphile molecules into nanofibers by electrostatic attraction. J Am Chem Soc 125:7146–7147. doi:10.1021/ja028215r
268. Xing X, Wang Y, Li B (2008) Nanofibers drawing and nanodevices assembly in poly(trimethylene terephthalate). Opt Express 16:10815. doi:10.1364/OE.16.010815
269. McKenzie JL, Waid MC, Shi R, Webster TJ (2004) Decreased functions of astrocytes on carbon nanofiber materials. Biomaterials 25:1309–1317. doi:10.1016/j.biomaterials.2003.08.006
270. Schnell E, Klinkhammer K, Balzer S et al (2007) Guidance of glial cell migration and axonal growth on electrospun nanofibers of poly-epsilon-caprolactone and a collagen/poly-epsilon-caprolactone blend. Biomaterials 28:3012–3025. doi:10.1016/j.biomaterials.2007.03.009
271. Yang F, Murugan R, Wang S, Ramakrishna S (2005) Electrospinning of nano/micro scale poly(L-lactic acid) aligned fibers and their potential in neural tissue engineering. Biomaterials 26:2603–2610. doi:10.1016/j.biomaterials.2004.06.051
272. Christopherson GT, Song H, Mao H-Q (2009) The influence of fiber diameter of electrospun substrates on neural stem cell differentiation and proliferation. Biomaterials 30:556–564. doi:10.1016/j.biomaterials.2008.10.004
273. Corey JM, Lin DY, Mycek KB et al (2007) Aligned electrospun nanofibers specify the direction of dorsal root ganglia neurite growth. J Biomed Mater Res A 83:636–645. doi:10.1002/jbm.a.31285
274. Mahairaki V, Lim SH, Christopherson GT et al (2011) Nanofiber matrices promote the neuronal differentiation of human embryonic stem cell-derived neural precursors in vitro. Tissue Eng Part A 17:855–863. doi:10.1089/ten.TEA.2010.0377
275. Lim SH, Mao H-Q (2009) Electrospun scaffolds for stem cell engineering. Adv Drug Deliv Rev 61:1084–1096. doi:10.1016/j.addr.2009.07.011
276. Lee J, Lee YJ, Cho H, Shin H (2014) Guidance of in vitro migration of human mesenchymal stem cells and in vivo guided bone regeneration using aligned electrospun fibers. Tissue Eng Part A 20:2031–2042. doi:10.1089/ten.tea.2013.0282
277. Wang J, Ye R, Wei Y et al (2012) The effects of electrospun TSF nanofiber diameter and alignment on neuronal differentiation of human embryonic stem cells. J Biomed Mater Res Part A 100A:632–645. doi:10.1002/jbm.a.33291
278. Lee MR, Kwon KW, Jung H et al (2010) Direct differentiation of human embryonic stem cells into selective neurons on nanoscale ridge/groove pattern arrays. Biomaterials 31:4360–4366. doi:10.1016/j.biomaterials.2010.02.012
279. Xie J, Willerth SM, Li X et al (2009) The differentiation of embryonic stem cells seeded on electrospun nanofibers into neural lineages. Biomaterials 30:354–362. doi:10.1016/j.biomaterials.2008.09.046

280. Recknor JB, Sakaguchi DS, Mallapragada SK (2006) Directed growth and selective differentiation of neural progenitor cells on micropatterned polymer substrates. Biomaterials 27:4098–4108. doi:10.1016/j.biomaterials.2006.03.029
281. Hu X, Liu S, Zhou G et al (2014) Electrospinning of polymeric nanofibers for drug delivery applications. J Control Release 185:12–21. doi:10.1016/j.jconrel.2014.04.018
282. Ignatious F, Baldoni J (2002) US Patent Appl. No 10/181,640
283. Sun Z, Zussman E, Yarin AL et al (2003) Compound core–shell polymer nanofibers by co-electrospinning. Adv Mater 15:1929–1932. doi:10.1002/adma.200305136
284. He C-L, Huang Z-M, Han X-J (2009) Fabrication of drug-loaded electrospun aligned fibrous threads for suture applications. J Biomed Mater Res A 89:80–95. doi:10.1002/jbm.a.32004
285. Qi M, Li X, Yang Y, Zhou S (2008) Electrospun fibers of acid-labile biodegradable polymers containing ortho ester groups for controlled release of paracetamol. Eur J Pharm Biopharm 70:445–452
286. Chew SY, Wen J, Yim EKF, Leong KW (2005) Sustained release of proteins from electrospun biodegradable fibers. Biomacromolecules 6:2017–2024. doi:10.1021/bm0501149
287. Kenawy E-R, Bowlin GL, Mansfield K et al (2002) Release of tetracycline hydrochloride from electrospun poly(ethylene-co-vinylacetate), poly(lactic acid), and a blend. J Control Release 81:57–64. doi:10.1016/S0168-3659(02)00041-X
288. Zeng J, Xu X, Chen X et al (2003) Biodegradable electrospun fibers for drug delivery. J Control Release 92:227–231. doi:10.1016/S0168-3659(03)00372-9
289. Yang Y, Li X, Qi M et al (2008) Release pattern and structural integrity of lysozyme encapsulated in core-sheath structured poly(DL-lactide) ultrafine fibers prepared by emulsion electrospinning. Eur J Pharm Biopharm Off J 69:106–116. doi:10.1016/j.ejpb.2007.10.016
290. Yang Y, Li X, Cui W et al (2008) Structural stability and release profiles of proteins from core-shell poly (DL-lactide) ultrafine fibers prepared by emulsion electrospinning. J Biomed Mater Res A 86:374–385. doi:10.1002/jbm.a.31595
291. Cui W, Li X, Zhu X et al (2006) Investigation of drug release and matrix degradation of electrospun poly(DL-lactide) fibers with paracetanol inoculation. Biomacromolecules 7:1623–1629. doi:10.1021/bm060057z
292. Cui W, Qi M, Li X et al (2008) Electrospun fibers of acid-labile biodegradable polymers with acetal groups as potential drug carriers. Int J Pharm 361:47–55. doi:10.1016/j.ijpharm.2008.05.005
293. Yuan Z, Zhao J, Zhu W et al (2014) Ibuprofen-loaded electrospun fibrous scaffold doped with sodium bicarbonate for responsively inhibiting inflammation and promoting muscle wound healing in vivo. Biomater Sci 2:502–511. doi:10.1039/C3BM60198F
294. Panyam J, Labhasetwar V (2003) Biodegradable nanoparticles for drug and gene delivery to cells and tissue. Adv Drug Deliv Rev 55:329–347. doi:10.1016/S0169-409X(02)00228-4
295. Teekamp N, Duque LF, Frijlink HW et al (2015) Production methods and stabilization strategies for polymer-based nanoparticles and microparticles for parenteral delivery of peptides and proteins. Expert Opin Drug Deliv 12:1311–1331. doi:10.1517/17425247.2015.1003807
296. Felice B, Prabhakaran MP, Rodríguez AP, Ramakrishna S (2014) Drug delivery vehicles on a nano-engineering perspective. Mater Sci Eng C Mater Biol Appl 41:178–195. doi:10.1016/j.msec.2014.04.049
297. Anderson JM, Shive MS (2012) Biodegradation and biocompatibility of PLA and PLGA microspheres. Adv Drug Deliv Rev 64:72–82. doi:10.1016/j.addr.2012.09.004
298. Matsusue Y, Hanafusa S, Yamamuro T et al (1995) Tissue Reaction of Bioabsorbable Ultra High Strength Poly (L-Lactide) Rod: A Long-Term Study in Rabbits. Clin Orthop Relat Res 317
299. Lin SY, Chen KS, Teng HH, Li MJ (2008) In vitro degradation and dissolution behaviours of microspheres prepared by three low molecular weight polyesters. J Microencapsul 17:577–586. doi:10.1080/026520400417630
300. Bracken MB, Shepard MJ, Collins WF et al (1990) A randomized, controlled trial of methylprednisolone or naloxone in the treatment of acute spinal-cord injury. N Engl J Med 322:1405–1411. doi:10.1056/NEJM199005173222001

301. Bydon M, Lin J, Macki M et al (2014) The current role of steroids in acute spinal cord injury. World Neurosurg 82:848–854. doi:10.1016/j.wneu.2013.02.062
302. Kim Y, Caldwell J-M, Bellamkonda RV (2009) Nanoparticle-mediated local delivery of Methylprednisolone after spinal cord injury. Biomaterials 30:2582–2590. doi:10.1016/j.biomaterials.2008.12.077
303. Cerqueira SR, Oliveira JM, Silva NA et al (2013) Microglia response and in vivo therapeutic potential of methylprednisolone-loaded dendrimer nanoparticles in spinal cord injury. Small 9:738–749. doi:10.1002/smll.201201888
304. Jain NK, Mishra V, Mehra NK (2013) Targeted drug delivery to macrophages. Expert Opin Drug Deliv 10:353–367. doi:10.1517/17425247.2013.751370
305. Hu W, Metselaar J, Ben L-H et al (2009) PEG minocycline-liposomes ameliorate CNS autoimmune disease. PLoS One 4, e4151. doi:10.1371/journal.pone.0004151
306. Wang Y-C, Wu Y-T, Huang H-Y et al (2008) Sustained intraspinal delivery of neurotrophic factor encapsulated in biodegradable nanoparticles following contusive spinal cord injury. Biomaterials 29:4546–4553. doi:10.1016/j.biomaterials.2008.07.050
307. Lu K-W, Chen Z-Y, Jin D-D et al (2002) Cationic liposome-mediated GDNF gene transfer after spinal cord injury. J Neurotrauma 19:1081–1090. doi:10.1089/089771502760341983
308. Hedley ML, Curley J, Urban R (1998) Microspheres containing plasmid-encoded antigens elicit cytotoxic T-cell responses. Nat Med 4:365–368. doi:10.1038/nm0398-365
309. Kim D, Schallert T, Browarak T et al (2001) Transplantation of Genetically Modified Fibroblasts Expressing BDNF in Adult Rats with a Subtotal Hemisection Improves Specific Motor and Sensory Functions. Neurorehabil Neural Repair 15:141–150. doi:10.1177/154596830101500207
310. Langer R (1990) New methods of drug delivery. Science (80-) 249:1527–1533.
311. Nie L, Zhang G, Hou R et al (2015) Controllable promotion of chondrocyte adhesion and growth on PVA hydrogels by controlled release of TGF-β1 from porous PLGA microspheres. Colloids Surf B Biointerfaces 125:51–57. doi:10.1016/j.colsurfb.2014.11.010
312. Cao X, Shoichet MS (1999) Delivering neuroactive molecules from biodegradable microspheres for application in central nervous system disorders. Biomaterials 20:329–339. doi:10.1016/S0142-9612(98)00172-0
313. Benoit J-P, Faisant N, Venier-Julienne M-C, Menei P (2000) Development of microspheres for neurological disorders: from basics to clinical applications. J Control Release 65:285–296. doi:10.1016/S0168-3659(99)00250-3
314. Burdick JA, Ward M, Liang E et al (2006) Stimulation of neurite outgrowth by neurotrophins delivered from degradable hydrogels. Biomaterials 27:452–459. doi:10.1016/j.biomaterials.2005.06.034
315. Kim H, Zahir T, Tator CH, Shoichet MS (2011) Effects of dibutyryl cyclic-AMP on survival and neuronal differentiation of neural stem/progenitor cells transplanted into spinal cord injured rats. PLoS One 6, e21744. doi:10.1371/journal.pone.0021744
316. Ashton RS, Banerjee A, Punyani S et al (2007) Scaffolds based on degradable alginate hydrogels and poly(lactide-co-glycolide) microspheres for stem cell culture. Biomaterials 28:5518–5525. doi:10.1016/j.biomaterials.2007.08.038
317. Prang P, Müller R, Eljaouhari A et al (2006) The promotion of oriented axonal regrowth in the injured spinal cord by alginate-based anisotropic capillary hydrogels. Biomaterials 27:3560–3569. doi:10.1016/j.biomaterials.2006.01.053
318. Read T-A, Stensvaag V, Vindenes H et al (1999) Cells encapsulated in alginate: a potential system for delivery of recombinant proteins to malignant brain tumours. Int J Dev Neurosci 17:653–663. doi:10.1016/S0736-5748(99)00052-0
319. Tobias CA, Han SSW, Shumsky JS et al (2005) Alginate encapsulated BDNF-producing fibroblast grafts permit recovery of function after spinal cord injury in the absence of immune suppression. 22:138–156
320. Pêgo AP, Kubinova S, Cizkova D et al (2012) Regenerative medicine for the treatment of spinal cord injury: more than just promises? J Cell Mol Med 16:2564–2582. doi:10.1111/j.1582-4934.2012.01603.x
321. Gennai S, Monsel A, Hao Q et al (2015) Cell-based therapy for traumatic brain injury. Br J Anaesth 115:203–212. doi:10.1093/bja/aev229

322. Emerich DF, Winn SR, Harper J et al (1994) Implants of polymer-encapsulated human NGF-secreting cells in the nonhuman primate: rescue and sprouting of degenerating cholinergic basal forebrain neurons. J Comp Neurol 349:148–164. doi:10.1002/cne.903490110
323. Liu YY (1999) Transplants of fibroblasts genetically modified to express BDNF promote regeneration of adult rat rubrospinal axons and recovery of forelimb function. J Neurosci 19:4370–4387
324. Kabu S, Gao Y, Kwon BK, Labhasetwar V (2015) Drug delivery, cell-based therapies, and tissue engineering approaches for spinal cord injury. J Control Release 219:141–154. doi:10.1016/j.jconrel.2015.08.060
325. Yang H, Liu C-C, Wang C-Y et al (2015) Therapeutical strategies for spinal cord injury and a promising autologous astrocyte-based therapy using efficient reprogramming techniques. Mol Neurobiol. doi:10.1007/s12035-015-9157-7
326. Ben-David U, Benvenisty N (2011) The tumorigenicity of human embryonic and induced pluripotent stem cells. Nat Rev Cancer 11:268–277. doi:10.1038/nrc3034
327. Réthelyi M, Horváth-Oszwald E, Boros C (2008) Caudal end of the rat spinal dorsal horn. Neurosci Lett 445:153–157. doi:10.1016/j.neulet.2008.08.070
328. Boros C, Lukácsi E, Horváth-Oszwald E, Réthelyi M (2008) Neurochemical architecture of the filum terminale in the rat. Brain Res 1209:105–114. doi:10.1016/j.brainres.2008.02.029
329. Varghese M, Olstorn H, Murrell W, Langmoen IA (2010) Exploring atypical locations of mammalian neural stem cells: the human filum terminale. Arch Ital Biol 148:85–94. doi:10.4449/aib.v148i2.1134
330. Arvidsson L, Fagerlund M, Jaff N et al (2011) Distribution and characterization of progenitor cells within the human filum terminale. PLoS One 6, e27393. doi:10.1371/journal.pone.0027393
331. Jha RM, Chrenek R, Magnotti LM, Cardozo DL (2013) The isolation, differentiation, and survival in vivo of multipotent cells from the postnatal rat filum terminale. PLoS One 8, e65974. doi:10.1371/journal.pone.0065974
332. Slovinska L, Szekiova E, Blasko J et al (2015) Comparison of dynamic behavior and maturation of neural multipotent cells derived from different spinal cord developmental stages: An in vitro study. Acta Neurobiol Exp (Wars) 75:107–114
333. Chen G, Hu YR, Wan H et al (2010) Functional recovery following traumatic spinal cord injury mediated by a unique polymer scaffold seeded with neural stem cells and Schwann cells. Chin Med J 123:2424–2431
334. Parr AM, Kulbatski I, Zahir T et al (2008) Transplanted adult spinal cord-derived neural stem/progenitor cells promote early functional recovery after rat spinal cord injury. Neuroscience 155:760–770. doi:10.1016/j.neuroscience.2008.05.042
335. Pfeifer K, Vroemen M, Caioni M et al (2006) Autologous adult rodent neural progenitor cell transplantation represents a feasible strategy to promote structural repair in the chronically injured spinal cord. Regen Med 1:255–266. doi:10.2217/17460751.1.2.255
336. Teng YD, Lavik EB, Qu X et al (2002) Functional recovery following traumatic spinal cord injury mediated by a unique polymer scaffold seeded with neural stem cells. Proc Natl Acad Sci U S A 99:3024–3029. doi:10.1073/pnas.052678899
337. Nemati SN, Jabbari R, Hajinasrollah M et al (2014) Transplantation of adult monkey neural stem cells into a contusion spinal cord injury model in rhesus macaque monkeys. Cell J 16:117–130
338. Lladó J, Haenggeli C, Maragakis NJ et al (2004) Neural stem cells protect against glutamate-induced excitotoxicity and promote survival of injured motor neurons through the secretion of neurotrophic factors. Mol Cell Neurosci 27:322–331. doi:10.1016/j.mcn.2004.07.010
339. Lu P, Jones L, Snyder E, Tuszynski M (2003) Neural stem cells constitutively secrete neurotrophic factors and promote extensive host axonal growth after spinal cord injury. Exp Neurol 181:115–129. doi:10.1016/S0014-4886(03)00037-2
340. Anderson L, Caldwell MA (2007) Human neural progenitor cell transplants into the subthalamic nucleus lead to functional recovery in a rat model of Parkinson's disease. Neurobiol Dis 27:133–140. doi:10.1016/j.nbd.2007.03.015

341. Ebert AD, Beres AJ, Barber AE, Svendsen CN (2008) Human neural progenitor cells over-expressing IGF-1 protect dopamine neurons and restore function in a rat model of Parkinson's disease. Exp Neurol 209:213–223. doi:10.1016/j.expneurol.2007.09.022
342. Shear DA, Tate MC, Archer DR et al (2004) Neural progenitor cell transplants promote long-term functional recovery after traumatic brain injury. Brain Res 1026:11–22. doi:10.1016/j.brainres.2004.07.087
343. Koutsoudaki PN, Papastefanaki F, Stamatakis A et al (2015) Neural stem/progenitor cells differentiate into oligodendrocytes, reduce inflammation, and ameliorate learning deficits after transplantation in a mouse model of traumatic brain injury. Glia. doi:10.1002/glia.22959
344. Tsukamoto A, Uchida N, Capela A et al (2013) Clinical translation of human neural stem cells. Stem Cell Res Ther 4:102. doi:10.1186/scrt313
345. Doetsch F, Caillé I, Lim DA et al (1999) Subventricular zone astrocytes are neural stem cells in the adult mammalian brain. Cell 97:703–716. doi:10.1016/S0092-8674(00)80783-7
346. Morshead C, Craig C, van der Kooy D (1998) In vivo clonal analyses reveal the properties of endogenous neural stem cell proliferation in the adult mammalian forebrain. Development 125:2251–2261
347. Wright LS, Prowse KR, Wallace K et al (2006) Human progenitor cells isolated from the developing cortex undergo decreased neurogenesis and eventual senescence following expansion in vitro. Exp Cell Res 312:2107–2120. doi:10.1016/j.yexcr.2006.03.012
348. Neri M, Maderna C, Ferrari D et al (2010) Robust generation of oligodendrocyte progenitors from human neural stem cells and engraftment in experimental demyelination models in mice. PLoS One 5, e10145. doi:10.1371/journal.pone.0010145
349. NeuralStem Inc. Human Spinal Cord Derived Neural Stem Cell Transplantation for the Treatment of Amyotrophic Lateral Sclerosis (ALS) (2014) In: ClinicalTrials.gov [Internet]. National Library of Medicine, Bethesda. https://clinicaltrials.gov. Accessed 11 Feb 2016
350. Trounson A, Thakar RG, Lomax G, Gibbons D (2011) Clinical trials for stem cell therapies. BMC Med 9:52. doi:10.1186/1741-7015-9-52
351. Ong WY, Levine JM (1999) A light and electron microscopic study of NG2 chondroitin sulfate proteoglycan-positive oligodendrocyte precursor cells in the normal and kainate-lesioned rat hippocampus. Neuroscience 92:83–95. doi:10.1016/S0306-4522(98)00751-9
352. Bergles DE, Jahr CE (1997) Synaptic Activation of Glutamate Transporters in Hippocampal Astrocytes. Neuron 19:1297–1308. doi:10.1016/S0896-6273(00)80420-1
353. Windrem MS, Nunes MC, Rashbaum WK et al (2004) Fetal and adult human oligodendrocyte progenitor cell isolates myelinate the congenitally dysmyelinated brain. Nat Med 10:93–97. doi:10.1038/nm974
354. Cavazzin C, Ferrari D, Facchetti F et al (2006) Unique expression and localization of aquaporin-4 and aquaporin-9 in murine and human neural stem cells and in their glial progeny. Glia 53:167–181. doi:10.1002/glia.20256
355. Windrem MS, Roy NS, Wang J et al (2002) Progenitor cells derived from the adult human subcortical white matter disperse and differentiate as oligodendrocytes within demyelinated lesions of the rat brain. J Neurosci Res 69:966–975. doi:10.1002/jnr.10397
356. Webber DJ, van Blitterswijk M, Chandran S (2009) Neuroprotective effect of oligodendrocyte precursor cell transplantation in a long-term model of periventricular leukomalacia. Am J Pathol 175:2332–2342. doi:10.2353/ajpath.2009.090051
357. Cao Q, He Q, Wang Y et al (2010) Transplantation of ciliary neurotrophic factor-expressing adult oligodendrocyte precursor cells promotes remyelination and functional recovery after spinal cord injury. J Neurosci 30:2989–3001. doi:10.1523/JNEUROSCI.3174-09.2010
358. Nistor GI, Totoiu MO, Haque N et al (2005) Human embryonic stem cells differentiate into oligodendrocytes in high purity and myelinate after spinal cord transplantation. Glia 49:385–396. doi:10.1002/glia.20127
359. Keirstead HS, Nistor G, Bernal G et al (2005) Human embryonic stem cell-derived oligodendrocyte progenitor cell transplants remyelinate and restore locomotion after spinal cord injury. J Neurosci 25:4694–4705. doi:10.1523/JNEUROSCI.0311-05.2005

360. Wilkins A, Majed H, Layfield R et al (2003) Oligodendrocytes Promote Neuronal Survival and Axonal Length by Distinct Intracellular Mechanisms: A Novel Role for Oligodendrocyte-Derived Glial Cell Line-Derived Neurotrophic Factor. J Neurosci 23:4967–4974
361. Erceg S, Ronaghi M, Oria M et al (2010) Transplanted oligodendrocytes and motoneuron progenitors generated from human embryonic stem cells promote locomotor recovery after spinal cord transection. Stem Cells 28:1541–1549. doi:10.1002/stem.489
362. Chen L-X, Ma S-M, Zhang P et al (2015) Neuroprotective effects of oligodendrocyte progenitor cell transplantation in premature rat brain following hypoxic-ischemic injury. PLoS One 10, e0115997. doi:10.1371/journal.pone.0115997
363. Mehrabi S, Eftekhari S, Moradi F et al (2013) Cell therapy in spinal cord injury: a mini-reivew. Basic Clin Neurosci 4:172–176
364. Xu XM, Guénard V, Kleitman N, Bunge MB (1995) Axonal regeneration into Schwann cell-seeded guidance channels grafted into transected adult rat spinal cord. J Comp Neurol 351:145–160. doi:10.1002/cne.903510113
365. Chen A, Xu XM, Kleitman N, Bunge MB (1996) Methylprednisolone administration improves axonal regeneration into Schwann cell grafts in transected adult rat thoracic spinal cord. Exp Neurol 138:261–276. doi:10.1006/exnr.1996.0065
366. Pinzon A, Calancie B, Oudega M, Noga BR (2001) Conduction of impulses by axons regenerated in a Schwann cell graft in the transected adult rat thoracic spinal cord. J Neurosci Res 64:533–541. doi:10.1002/jnr.1105
367. Guest JD, Hiester ED, Bunge RP (2005) Demyelination and Schwann cell responses adjacent to injury epicenter cavities following chronic human spinal cord injury. Exp Neurol 192:384–393. doi:10.1016/j.expneurol.2004.11.033
368. Pearse DD, Sanchez AR, Pereira FC et al (2007) Transplantation of Schwann cells and/or olfactory ensheathing glia into the contused spinal cord: Survival, migration, axon association, and functional recovery. Glia 55:976–1000. doi:10.1002/glia.20490
369. Takami T, Oudega M, Bates ML et al (2002) Schwann cell but not olfactory ensheathing glia transplants improve hindlimb locomotor performance in the moderately contused adult rat thoracic spinal cord. J Neurosci 22:6670–6681
370. Kamada T, Koda M, Dezawa M et al (2011) Transplantation of human bone marrow stromal cell-derived Schwann cells reduces cystic cavity and promotes functional recovery after contusion injury of adult rat spinal cord. Neuropathology 31:48–58. doi:10.1111/j.1440-1789.2010.01130.x
371. Bhatheja K, Field J (2006) Schwann cells: origins and role in axonal maintenance and regeneration. Int J Biochem Cell Biol 38:1995–1999. doi:10.1016/j.biocel.2006.05.007
372. Olson HE, Rooney GE, Gross L et al (2009) Neural Stem Cell– and Schwann Cell–Loaded Biodegradable Polymer Scaffolds Support Axonal Regeneration in the Transected Spinal Cord. Tissue Eng Part A 15:1797–1805. doi:10.1089/ten.tea.2008.0364
373. Afshari FT, Kwok JC, Fawcett JW (2010) Astrocyte-produced ephrins inhibit schwann cell migration via VAV2 signaling. J Neurosci 30:4246–4255. doi:10.1523/JNEUROSCI.3351-09.2010
374. Sekiya I, Larson BL, Smith JR et al (2002) Expansion of human adult stem cells from bone marrow stroma: conditions that maximize the yields of early progenitors and evaluate their quality. Stem Cells 20:530–541. doi:10.1634/stemcells.20-6-530
375. Dasari VR, Veeravalli KK, Dinh DH (2014) Mesenchymal stem cells in the treatment of spinal cord injuries: a review. World J Stem Cells 6:120–133. doi:10.4252/wjsc.v6.i2.120
376. Kim S, Honmou O, Kato K et al (2006) Neural differentiation potential of peripheral blood- and bone-marrow-derived precursor cells. Brain Res 1123:27–33. doi:10.1016/j.brainres.2006.09.044
377. Xu H, Miki K, Ishibashi S et al (2010) Transplantation of neuronal cells induced from human mesenchymal stem cells improves neurological functions after stroke without cell fusion. J Neurosci Res 88:3598–3609. doi:10.1002/jnr.22501
378. Cho S-R, Kim YR, Kang H-S et al (2009) Functional recovery after the transplantation of neurally differentiated mesenchymal stem cells derived from bone barrow in a rat model of spinal cord injury. Cell Transplant 18:1359–1368. doi:10.3727/096368909X475329

379. Osaka M, Honmou O, Murakami T et al (2010) Intravenous administration of mesenchymal stem cells derived from bone marrow after contusive spinal cord injury improves functional outcome. Brain Res 1343:226–235. doi:10.1016/j.brainres.2010.05.011
380. Karussis D, Kassis I, Kurkalli BGS, Slavin S (2008) Immunomodulation and neuroprotection with mesenchymal bone marrow stem cells (MSCs): a proposed treatment for multiple sclerosis and other neuroimmunological/neurodegenerative diseases. J Neurol Sci 265:131–135. doi:10.1016/j.jns.2007.05.005
381. Titomanlio L, Kavelaars A, Dalous J et al (2011) Stem cell therapy for neonatal brain injury: Perspectives and Challenges. Ann Neurol 70:698–712. doi:10.1002/ana.22518
382. Syková E, Jendelová P, Urdzíková L, et al. Bone marrow stem cells and polymer hydrogels–two strategies for spinal cord injury repair. Cell Mol Neurobiol 26:1113–1129. doi:10.1007/s10571-006-9007-2
383. van Velthoven CTJ, Kavelaars A, van Bel F, Heijnen CJ (2010) Repeated mesenchymal stem cell treatment after neonatal hypoxia-ischemia has distinct effects on formation and maturation of new neurons and oligodendrocytes leading to restoration of damage, corticospinal motor tract activity, and sensorimotor function. J Neurosci 30:9603–9611. doi:10.1523/JNEUROSCI.1835-10.2010
384. Mahmood A, Lu D, Chopp M (2004) Marrow stromal cell transplantation after traumatic brain injury promotes cellular proliferation within the brain. Neurosurgery 55
385. Okazaki T, Magaki T, Takeda M et al (2008) Intravenous administration of bone marrow stromal cells increases survivin and Bcl-2 protein expression and improves sensorimotor function following ischemia in rats. Neurosci Lett 430:109–114. doi:10.1016/j.neulet.2007.10.046
386. Quertainmont R, Cantinieaux D, Botman O et al (2012) Mesenchymal stem cell graft improves recovery after spinal cord injury in adult rats through neurotrophic and pro-angiogenic actions. PLoS One 7, e39500. doi:10.1371/journal.pone.0039500
387. Ide C, Nakai Y, Nakano N et al (2010) Bone marrow stromal cell transplantation for treatment of sub-acute spinal cord injury in the rat. Brain Res 1332:32–47. doi:10.1016/j.brainres.2010.03.043
388. Franssen EHP, de Bree FM, Verhaagen J (2007) Olfactory ensheathing glia: their contribution to primary olfactory nervous system regeneration and their regenerative potential following transplantation into the injured spinal cord. Brain Res Rev 56:236–258. doi:10.1016/j.brainresrev.2007.07.013
389. Raisman G, Li Y (2007) Repair of neural pathways by olfactory ensheathing cells. Nat Rev Neurosci 8:312–319. doi:10.1038/nrn2099
390. Ramón-Cueto A, Plant GW, Avila J, Bunge MB (1998) Long-distance axonal regeneration in the transected adult rat spinal cord is promoted by olfactory ensheathing glia transplants. J Neurosci 18:3803–3815
391. Li Y (1997) Repair of adult rat corticospinal tract by transplants of olfactory ensheathing cells. Science 277:2000–2002. doi:10.1126/science.277.5334.2000
392. Fouad K, Schnell L, Bunge MB et al (2005) Combining Schwann cell bridges and olfactory-ensheathing glia grafts with chondroitinase promotes locomotor recovery after complete transection of the spinal cord. J Neurosci 25:1169–1178. doi:10.1523/JNEUROSCI.3562-04.2005
393. Hospital CMU Implantation of olfactor ensheathing cells (OECs) (2011) In: ClinicalTrials.gov [Internet]. National Library of Medicine, Bethesda. https://clinicaltrials.gov/ct2/show/NCT02034669. Accessed 27 Jan 2016
394. Wroclaw Medical University Transplantation of Autologous Olfactor Ensheathing Cells in Complete Human Spinal Cord Injury (2010) In: ClinicalTrials.gov [Internet]. National Library of Medicine, Bethesda. https://clinicaltrials.gov/ct2/show/NCT0123. Accessed 27 Jan 2016
395. Leu S, Lin Y-C, Yuen C-M et al (2010) Adipose-derived mesenchymal stem cells markedly attenuate brain infarct size and improve neurological function in rats. J Transl Med 8:63. doi:10.1186/1479-5876-8-63
396. Ra JC, Shin IS, Kim SH et al (2011) Safety of intravenous infusion of human adipose tissue-derived mesenchymal stem cells in animals and humans. Stem Cells Dev 20:1297–1308. doi:10.1089/scd.2010.0466

397. Yan Y, Ma T, Gong K et al (2014) Adipose-derived mesenchymal stem cell transplantation promotes adult neurogenesis in the brains of Alzheimer's disease mice. Neural Regen Res 9:798–805. doi:10.4103/1673-5374.131596
398. Kang SK, Jun ES, Bae YC, Jung JS (2003) Interactions between human adipose stromal cells and mouse neural stem cells in vitro. Brain Res Dev Brain Res 145:141–149
399. Oh JS, Kim KN, An SS et al (2011) Cotransplantation of mouse neural stem cells (mNSCs) with adipose tissue-derived mesenchymal stem cells improves mNSC survival in a rat spinal cord injury model. Cell Transplant 20:837–849. doi:10.3727/096368910X539083
400. Oh JS, Park IS, Kim KN et al (2012) Transplantation of an adipose stem cell cluster in a spinal cord injury. Neuroreport 23
401. Kolar MK, Kingham PJ, Novikova LN et al (2014) The therapeutic effects of human adipose-derived stem cells in a rat cervical spinal cord injury model. Stem Cells Dev 23:1659–1674. doi:10.1089/scd.2013.0416
402. Zhou Z, Chen Y, Zhang H et al (2013) Comparison of mesenchymal stromal cells from human bone marrow and adipose tissue for the treatment of spinal cord injury. Cytotherapy 15:434–448. doi:10.1016/j.jcyt.2012.11.015
403. Reid AJ, Sun M, Wiberg M et al (2011) Nerve repair with adipose-derived stem cells protects dorsal root ganglia neurons from apoptosis. Neuroscience 199:515–522. doi:10.1016/j.neuroscience.2011.09.064
404. Zaminy A, Shokrgozar MA, Sadeghi Y et al (2013) Transplantation of Schwann Cells Differentiated from Adipose Stem Cells Improves Functional Recovery in Rat Spinal Cord Injury. Arch Iran Med 16:533–541
405. Biotechnology T, JSC Transplantation of Autologous Adipose Derived Stem Cells (ADSCs) in Spinal Cord Injury Treatment. In: ClinicalTrials.gov [Internet]. National Library of Medicine, Bethesda. https://clinicaltrials.gov/ct2/show/NCT0203669NLMIdent
406. Choy DKS, Nga VDW, Lim J et al (2013) Brain tissue interaction with three-dimensional, honeycomb polycaprolactone-based scaffolds designed for cranial reconstruction following traumatic brain injury. Tissue Eng Part A 19:2382–2389. doi:10.1089/ten.TEA.2012.0733
407. Sato Y, Nakanishi K, Hayakawa M et al (2008) Reduction of brain injury in neonatal hypoxic-ischemic rats by intracerebroventricular injection of neural stem/progenitor cells together with chondroitinase ABC. Reprod Sci 15:613–620. doi:10.1177/1933719108317299
408. Shi W, Nie D, Jin G et al (2012) BDNF blended chitosan scaffolds for human umbilical cord MSC transplants in traumatic brain injury therapy. Biomaterials 33:3119–3126. doi:10.1016/j.biomaterials.2012.01.009
409. Nakaguchi K, Jinnou H, Kaneko N et al (2012) Growth factors released from gelatin hydrogel microspheres increase new neurons in the adult mouse brain. Stem Cells Int 2012:915160. doi:10.1155/2012/915160
410. Thau-Zuchman O, Shohami E, Alexandrovich AG, Leker RR (2012) Combination of vascular endothelial and fibroblast growth factor 2 for induction of neurogenesis and angiogenesis after traumatic brain injury. J Mol Neurosci 47:166–172. doi:10.1007/s12031-012-9706-8
411. Johnson PJ, Tatara A, McCreedy DA et al (2010) Tissue-engineered fibrin scaffolds containing neural progenitors enhance functional recovery in a subacute model of SCI. Soft Matter 6:5127–5137. doi:10.1039/c0sm00173b
412. Wilems TS, Pardieck J, Iyer N, Sakiyama-Elbert SE (2015) Combination therapy of stem cell derived neural progenitors and drug delivery of anti-inhibitory molecules for spinal cord injury. Acta Biomater 28:23–32. doi:10.1016/j.actbio.2015.09.018
413. Xiong Y, Zhu J-X, Fang Z-Y et al (2012) Coseeded Schwann cells myelinate neurites from differentiated neural stem cells in neurotrophin-3-loaded PLGA carriers. Int J Nanomedicine 7:1977–1989. doi:10.2147/IJN.S30706
414. Xia L, Wan H, Hao S et al (2013) Co-transplantation of neural stem cells and Schwann cells within poly (L-lactic-co-glycolic acid) scaffolds facilitates axonal regeneration in hemisected rat spinal cord. Chin Med J (Engl) 126:909–917
415. Li Y, Yu Z (2014) Progress in application of the combination of neural stem cells and Schwann cells for nerve repairing. Lin Chung Er Bi Yan Hou Tou Jing Wai Ke Za Zhi 28:1006–1009

Index

J. Wilson-Rawls, K. Kusumi (eds.), *Innovations in Molecular Mechanisms and Tissue Engineering*, Stem Cell Biology and Regenerative Medicine,
DOI 10.1007/978-3-319-44996-8

MIX
Papier aus verantwortungsvollen Quellen
Paper from responsible sources
FSC® C105338

If you have any concerns about our products,
you can contact us on
ProductSafety@springernature.com

In case Publisher is established outside the EU,
the EU authorized representative is:
Springer Nature Customer Service Center GmbH
Europaplatz 3, 69115 Heidelberg, Germany

Printed by Libri Plureos GmbH
in Hamburg, Germany